SpringerBriefs in Energy
Energy Analysis

Series Editor: Charles A.S. Hall

For further volumes:
http://www.springer.com/series/10041

Charles A. S. Hall · Carlos A. Ramírez-Pascualli

The First Half of the Age of Oil

An Exploration of the Work of Colin Campbell and Jean Laherrère

 Springer

Charles A. S. Hall
State University of New York
College of Environmental Science &
    Forestry
Syracuse, NY, USA

Carlos A. Ramírez-Pascualli
State University of New York
College of Environmental Science &
    Forestry
Syracuse, NY, USA

ISSN 2191-5520   ISSN 2191-5539 (electronic)
ISBN 978-1-4614-6063-3   ISBN 978-1-4614-6064-0 (eBook)
DOI 10.1007/978-1-4614-6064-0
Springer New York Heidelberg Dordrecht London

Library of Congress Control Number: 2012951422

Printed on acid-free paper

Springer is part of Springer Science+Business Media (www.springer.com)

About the Authors

Professor Charles A. S. Hall is a systems ecologist with strong interests in energy flows in natural systems and human society. He received his Ph.D. from Dr. Howard Odum at the University of North Carolina at Chapel Hill in 1970. His work has involved streams, estuaries, and tropical forests but has focused increasingly on human-dominated ecosystems in the USA and Latin America. He is best known for developing the concept of EROI, or energy return on investment, as it relates to, e.g., migrating fish and obtaining oil and gas. Hall's latest focus has been on developing an alternative approach to economics called biophysical economics, an attempt to understand human economies from a biophysical rather than just social perspective. He recently coauthored *Energy and the Wealth of Nations: Understanding the Biophysical Economy* with economist Kent Klitgaard.

Carlos A. Ramírez-Pascualli is a Ph.D. student in environmental science at the State University of New York, College of Environmental Science and Forestry (SUNY-ESF), where he is doing research on the biophysical aspects of economic systems, specifically on the relation of oil production to the Mexican economy. He holds degrees from some of the leading institutions in Mexico and Latin America: M.Sc. in economics from El Colegio de México (COLMEX) and B.Sc. in Industrial Engineering from Universidad Nacional Autónoma de México (UNAM). Before entering the Ph.D. program at SUNY-ESF, he was part of the team that developed the main information system at the Federal Competition Commission in Mexico. Previously, he worked as researcher and teaching assistant in several microeconomic courses at COLMEX. In addition to his official degrees, he has studied statistics and enjoys reading as much philosophy as he can.

Preface

This book was originally conceived as a series of chapters written by Colin Campbell and Jean Laherrère themselves. But because of the difficulty in writing an entire volume for two men in their late 70s or early 80s, particularly given their still very busy schedules, it became a book about the immense contributions of Campbell and Laherrère—using their own words as much as possible. Carlos Ramírez-Pascualli took on the job of finding and melding their original words into a series of chapters. He undertook this project because, as a Ph.D. student in economics and the environment with a focus on petroleum, he believed this would be the best way to understand the longstanding debate between so-called "peakists and optimists," an issue that is crucial for his research. Charles Hall took on the job of polishing Carlos' usually excellent English, as well as helping in the overall structure of the book. Unless otherwise specified, the final words are, to the best of our knowledge, derived especially from Campbell, with many graphs and plots in Laherrère's preferred way to communicate scientific information.

A bibliography is provided at the end of each chapter. Whenever it was necessary to update or add more information with the text of Campbell and Laherrère, we include in-text citations which are associated with the list of "References" for each chapter. At the risk of being repetitive, we specified the units of measurement almost every time we cite figures, preferring to be repetitive than inaccurate. Colin Campbell and Jean Laherrère have approved the final version, but any omissions or misrepresentations are the sole responsibility of the authors of this volume.

Acknowledgments

We would like to thank Colin Campbell for permission to use his ideas, texts and materials in this book and Jean Laherrère for permission to reprint his graphs. To both of them, our sincere thanks for their patience and effort embodied in the comments to early versions of the book. We thank Ajay Gupta for his valuable edits, comments, and suggestions. Carlos Ramírez-Pascualli would like to acknowledge the financial support provided by the National Council on Science and Technology of Mexico (CONACYT) during the preparation of the book.

Contents

Acronyms

AIOC	Anglo-Iranian Oil Company
API	American Petroleum Institute
APOC	Anglo-Persian Oil Company
ARAMCO	Arabian-American Oil Company
ARCO	Atlantic Richfield Company
ASPO	Association for the Study of Peak Oil and Gas
BBC	British Broadcasting Corporation
BGR	Bundesanstalt für Geowissenschaften und Rohstoffe
BP	British Petroleum
CERA	Cambridge Energy Research Associates
CFP	Compagnie Francais des Pétroles
CNNOC	China National Offshore Oil Corporation
EIA	United States Energy Information Agency
ENI	Ente Nazionale Idrocarburi
EOR	Enhanced oil recovery
EROI	Energy return on (energy) investment
IEA	International Energy Agency
IHS	Information Handling Services
IUPAC	International Union of Pure and Applied Chemistry
MMS	Minerals Management Service
NGL	Natural gas liquids
ODAC	Oil Depletion Analysis Centre
OECD	Organisation for Economic Co-operation and Development
OPEC	Organization of the Petroleum Exporting Countries
PEMEX	Petróleos Mexicanos
SEC	United States Securities and Exchange Commission
USDOE	United States Department of Energy
USGS	United States Geological Survey
USL48	United States' Lower 48 states
USSR	Union of Soviet Socialist Republics
WEO	World Energy Outlook XH Extra heavy oil

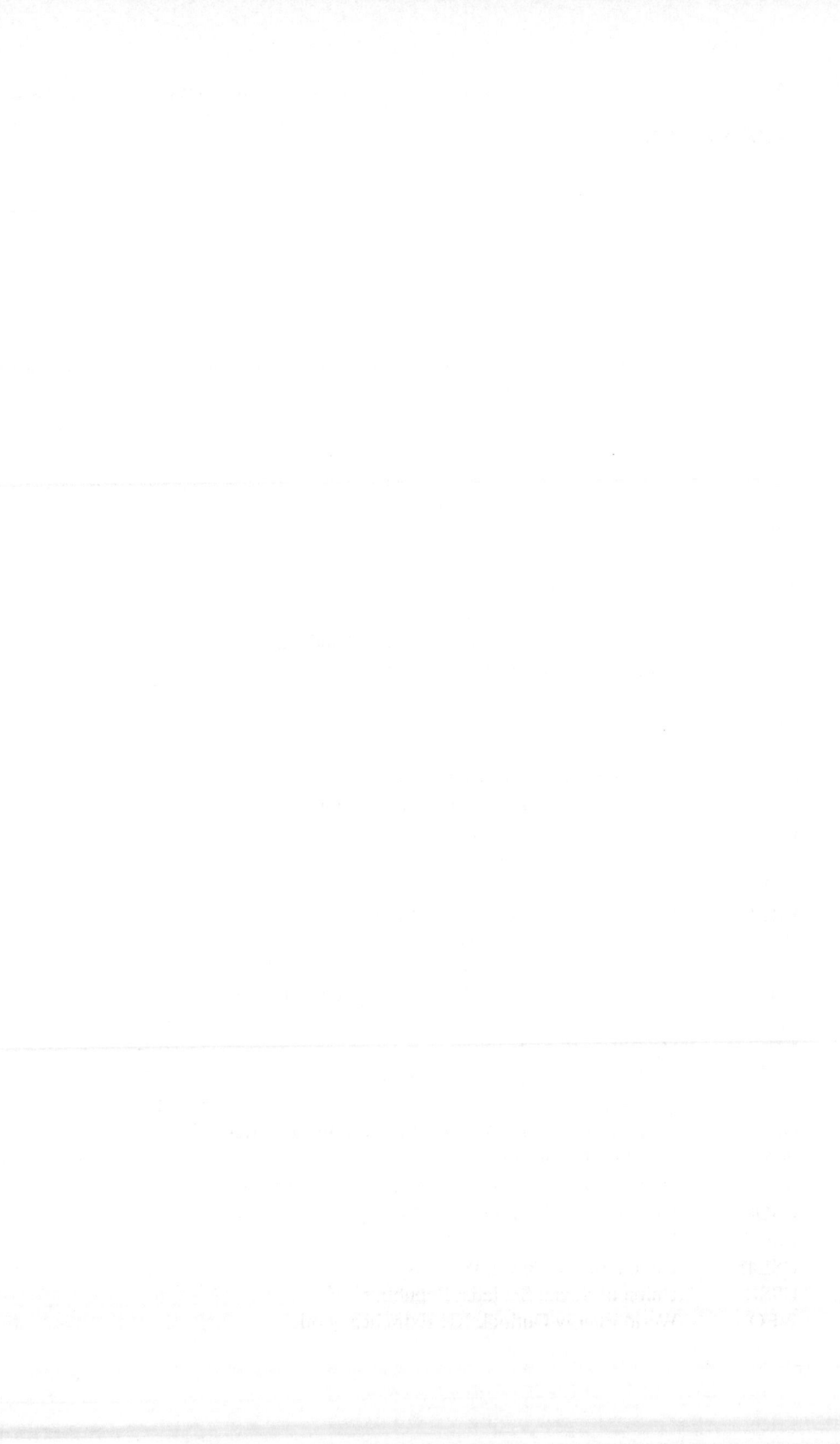

Chapter 1
Introduction

The efforts of Colin Campbell and Jean Laherrère have influenced the way in which we understand natural resources, in particular oil and natural gas. They have worked for more than 15 years to raise awareness about the implications of resource constraints for oil and gas production at the global level. Even though they are not the first analysts to study oil data and forecast a decline in global oil production, their work, especially the 1998 article titled "The End of Cheap Oil," represents an important milestone in the long-term debate about resource availability. The results presented in the article stand out from many other works due to their extensive and careful analysis. With a combined international experience of nearly 70 years inside the oil industry, in addition to their experience as external consultants after retirement, Campbell and Laherrère earned the trust of a growing community of analysts, scholars, and activists that assembled the Association for the Study of Peak Oil and Gas (ASPO), an international network of professionals from different backgrounds who are concerned with the depletion of the world's endowment of oil and gas and its possible impacts.

Campbell and Laherrère are not politicians, businessmen, or scholars, yet their ideas and analyses have had and will continue to have a significant impact in energy topics, business strategy, and geopolitical issues. Despite its importance, "The End of Cheap Oil" is only one piece in the vast intellectual production of both authors. They had been interested in this topic for a long time, so it was very difficult to choose what to include and what to leave out of a book like this. Colin Campbell has written two books and more than 150 papers about the oil industry, in which he has presented his views about the significance of oil for modern civilization. Jean Laherrère has also produced several dozens of articles and presentations, collecting, correcting, and plotting all the data he can get on energy, fossil fuels, and also other mineral resources. He has presented his results in numerous conferences and seminars; his graphs have been used by the International Energy Agency and the World Energy Council. Taken together, this combination of skills produces a body of knowledge that is worthy of discussion and enrichment with other points of view. Therefore, we believe that it is necessary to spread and discuss their work. We strongly recommend the interested reader to get their original works, many of which can be found online.

C.A.S. Hall and C.A. Ramírez-Pascualli, *The First Half of the Age of Oil: An Exploration of the Work of Colin Campbell and Jean Laherrère*, SpringerBriefs in Energy, DOI 10.1007/978-1-4614-6064-0_1, © Springer Science+Business Media New York 2013

Thus, this book was written as part of a dissemination effort aimed to synthesize and expand the views of Colin Campbell and Jean Laherrère in an organized and structured way. The text is based on documents authored by Campbell and Laherrère but updated with recent information coming from topics that range from the earth and life sciences to the social, economic, and geopolitical interactions at the global level, always trying to respect the meaning of the original text as much as possible. Even though we sympathize with and admire many of their ideas, as scholars, we think it is essential to address the weak points of their arguments for the sake of advancing the discussion. While we do not claim any credit for the insightful analyses of Campbell and Laherrère, we take full responsibility for the contents of this book, including any mistakes, misrepresentations, and omissions.

We have organized this book into three parts. The first, in which we discuss the evolution and history of oil and natural gas as prime resources for modern civilization, comprises three chapters. In Chap. 2, we unravel the connections between oil, money, and the industrial society in which we live. We analyze the geological formation of oil and natural gas, a process that involved a series of specific and irregular conditions in our planet's history. These fossil fuels, due to their unique physical characteristics, became the prime energy sources of the industry, influencing the social relations that unfolded upon the industrial mode of production. In Chap. 3, we explore those physical properties that make oil and gas so important today. But these characteristics have also implied limits, challenges, and conflicts for our societies. Hence, in Chap. 4 we provide a quick overview of the birth of the modern oil industry. Whether in North America, at the shores of the Caspian Sea, or in the Middle East, the oil industry was neither the result of the human capacity to "master nature" nor the logical consequence of "modern progress." Oil extraction was dominated by companies originally based in a few countries, either in North America or Europe. The major difference between both continents was that the American companies extracted their oil locally at first, while the Europeans had to bring the oil from abroad. Soon, oil became a strategic military resource for the Europeans, contributing to increase the tensions between Germany and England that would result in World War I, a struggle that started with horses and finished with tanks and air fighters. Hence, the most important factors that pushed petroleum to the level of importance it has had ever since were not the economic forces of supply and demand but rather the result of geology, geography and political conflict.

The second part, composed by three chapters as well, is dedicated to explain the ideas and techniques used to analyze oil production, in particular the models developed by M. King Hubbert and enhanced by Campbell and Laherrère for the world as a whole. Chapter 5 is a brief overview of the technique used by Hubbert in his influential lecture of 1956, titled "Nuclear Energy and the Fossil Fuels," where he predicted a peak in the production of oil in the US, considered only as the lower 48 states. Hubbert did not explore the prospects of Alaskan oil for reasons that we were not able to clarify, and his prediction has been criticized due to this fact. This prediction, however, needs to be understood as part of a larger argument; Hubbert was not trying to forecast the future production of fossil fuels *per se* but raising the issue of energy security at the national and international level in the context of the

1950s, a time when the oil industry appeared to be growing healthily and discussing geological constraints seemed to be a waste of time. Even though he was not the first to predict a production decline, he did so with the clear consciousness of launching a challenge to the insularity in the oil industry and in the energy sector at large. Today, this challenge seems to be more alive than ever.

In 1998, Campbell and Laherrère expanded the basic analysis of Hubbert using a large and reliable database for the world as a whole. As mentioned before, they published their results in "The End of Cheap Oil," an article that has more than a thousand recorded citations in different kinds of publications from several disciplines around the world. Since the database used by Campbell and Laherrère in 1998 is private, its information is no longer available for a similar analysis. Thus we present an updated version of the article, using data from the US Energy Information Administration and some recent graphs by Jean Laherrère in Chap. 6. In Chap. 7, we explore historical data recently updated and analyzed by Jean Laherrère and explain how it seems to be supporting the hypothesis of a world with nongrowing oil supplies.

The third part is an overview of the efforts undertaken by Campbell and Laherrère after the publication of "The End of Cheap Oil" and the discussions they have sustained directly or indirectly with analysts and official institutions. In Chap. 8, we make a quick account of the formation and explosion of ASPO, a network now presided over by Professor Kjell Aleklett, from Uppsala University, in Sweden, and also the Rimini Protocol, an effort initially promoted by Colin Campbell that seeks an international agreement on the issue of oil and gas depletion. Chap. 9 is a brief analysis of the internal dynamics of economic decision-making inside the oil industry and a sample of the arguments presented against the perspective of Campbell, Laherrère, and ASPO. Unfortunately, due to lack of space in this volume, we could not discuss the position of other influential institutions (e.g., OPEC, US DOE, and US EIA) or the economic theories that are usually invoked to belittle the importance of natural resources and peak oil. We tried to take the arguments from the other side and summarize and criticize them with all due rigor.

Chapter 2
Oil, Money, and Our Modern Civilization

The modern world runs on oil and money. Money has no intrinsic value, but it grants access to oil and the energy-intense products derived from oil—that is, a large part of the goods and services of the modern society. Oil and money are linked as cheap energy—mainly oil based—has fueled the economic prosperity of the past century. An appropriate term for the geological and the economic time in which we live in is the *oleocene*—the age, of oil. Although some have said that we live in an information age or a postindustrial age, it is clear that our life is based fundamentally on hydrocarbons. Just look around. Oil and gas are, however, finite natural resources that were formed only under very rare and special conditions in the geological past, which means that they are subject to depletion. Today for every gallon used one less remains: it is a simple concept to grasp. Think of a glass of beer (e.g., Guinness stout) in your hand after the bar has run out of beer. You take a sip and there is less in the glass (Fig. 2.1). Of course, you may find another open bar if you are lucky enough. Now think all the beer factories are out of business. Would you drink that beer in your hand fast as the previous ones? You could import your beer from bars abroad (remember all factories are closed), or go to other countries to have a drink, if you can afford it, of course. Understand this and you can begin to understand our basic situation with respect to oil. This is why our account of the modern economy must begin long before the advent of the international oil companies or the development of the stock exchange market where oil is bought and sold. This is a story about geological processes, and geologists have a different sense of time than most others. So for our story, we have to go back to the dawn of time. Then we can see the very special and almost impossible sequence of events that have led to today's industrialized human society.

2.1 Planet Earth

Scientific evidence suggests that the Solar System came into effect with a disturbance of an interstellar gas cloud—the solar nebula—about 5 billion years ago. One of the members of the system is planet Earth, which has evolved differently from the other

C.A.S. Hall and C.A. Ramírez-Pascualli, *The First Half of the Age of Oil: An Exploration of the Work of Colin Campbell and Jean Laherrère*, SpringerBriefs in Energy, DOI 10.1007/978-1-4614-6064-0_2, © Springer Science+Business Media New York 2013

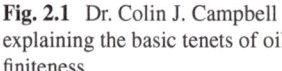
Fig. 2.1 Dr. Colin J. Campbell explaining the basic tenets of oil finiteness

known planets for some physical reasons that have not been clarified yet. The Earth has a molten core and a crust with segments—the tectonic plates—that have been moving around on the back of deep-seated convection currents, forming and shaping today's continents; entire mountain ranges have risen when these colossal plates collide with each other and massive lakes and even oceans appeared in the middle of the continents when the pieces moved apart. Immense energies coming from the core also caused volcanic eruptions, triggering a series of chemical reactions that led to the formation of water and air during the early history of the planet. These two ingredients formed a relatively thin and sensitive skin—the atmosphere—that is only a few tens of kilometers thick. Despite its thinness, the atmosphere came to have a huge impact on the future history of the planet, as the mountains lifted by the collision of the tectonic plates were exposed to rain and, hence, erosion. Rivers carried the debris down the mountain slopes to be deposited in lakes and seas. These deposits contained fine-grained material that formed a primeval ooze, which became home to the first forms of life.

The first microorganisms appeared on the planet about 3 billion years ago, followed by an explosion of life during the Cambrian period, around 550 million years ago. This explosion included, for the first time, various hard-shelled forms which remained preserved as fossils until our days. For example, the simple limpet (*Patella*, to give it its scientific name) has lived little changed for over 500 million years of geologically recorded history, having found a sustainable place in the environment, clinging to a rock as the waves washed over it. Other species were more adventurous, finding ways to exploit more effectively the particular niches in which they found themselves. Their numbers grew, as a genetic momentum perfected their adaptation. They were highly successful so long as their niche lasted, but the ever-changing environment surprised them from time to time in dramatic ways. Cataclysmic events, such as volcanic eruptions and changes in climate, affected these more sophisticated species more than the simpler forms; they did not manage to evolve backwards to the more sustainable simple stock from which they came and died out as victims of their very success. In this respect, the evolutionists did not get it exactly right when they explained evolution in terms of the "survival of the fittest." Evidently, the fittest over the short term were not the same as the fittest over the long term.

2.2   The Evolution of Fossil Fuels

Amidst the enormous geological and biological complexities and intricacies of the geological process occurring over the last 4 billion years, an extremely rare, almost trivial (from the perspective of the larger geological processes that were occurring) series of events resulted in the formation of oil and gas (and coal) that are so important to us and have generated so much of our present wealth.

It seems that in the entire history of the Earth there were only a few brief periods when a substantial amount of oil was formed, indicating that it took very specific conditions for its formation. According to the existing data, about 90% of the recoverable oil and gas reserves were generated during six intervals: (1) Silurian (9% of the world's reserves, 440–410 million years ago), (2) Upper Devonian—Tournaisian (8%, 375–350 million years ago), (3) Pennsylvanian—Lower Permian (8%, 320–290 million years ago), (4) Upper Jurassic (25%, 170–145 million years ago), (5) Middle Cretaceous (29%, 120–90 million years ago), and (6) Oligocene–Miocene (12.5%, 36–5 million years ago) (Klemme and Ulmishek 1991). According to these figures, oil formation represents <10% of the history of the planet. This information is being constantly updated and these numbers are far from being definitive, but they do give us a good idea of the issue. Why were the times that oil formed in significant amounts so rare?

First of all it needed to be a very warm period in the Earth's climate history. This was necessary to have very favorable conditions for the proliferation of small aquatic or marine plants (phytoplankton)—from which oil is formed—and for water bodies not to mix from top to bottom. Second it had to be during a period of time when there were important movements of the Earth's plates so that very deep lakes (or coastal regions) were formed. We can see similar conditions today occurring in the rift lakes of East Africa, where the coast of the continent is moving eastward, leaving a series of gorgeous, very deep lakes in its wake. Because these lakes are in the tropics, the surface of the water is very warm, which both favors the growth of algae and insures that the colder bottoms cannot be mixed with the surface. Since there is no mixing in the bottom, there is no oxygen transfer from the atmosphere (the bottom of such lakes has no oxygen). Similar conditions may have occurred during periods of global warming. This means that any algae that sink will not be oxidized to carbon dioxide but rather

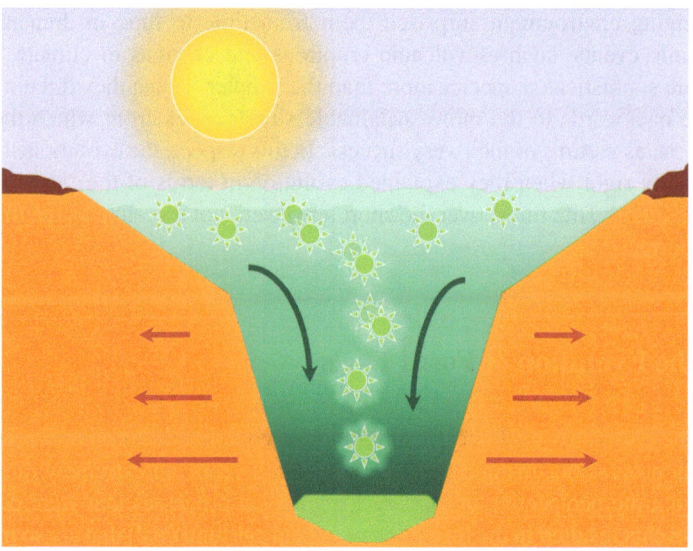

Fig. 2.2 Initial process in the formation of oil and gas reservoirs. Increased photosynthesis in non-mixing aquatic environments, likely due to global warming, during the period of formation of deep lakes or coastal areas, followed by the deposition of algae in an anoxic environment; modified and published with permission of Colin Campbell

will be attacked by anaerobic bacteria and accumulate as organic debris over thousands or millions of years (Fig. 2.2).

This is the beginning of the formation of oil, but there are further considerations. It is necessary to have a rainy climate—likely, as a consequence of a climate change—so that rivers bring massive amounts of silt and sand, covering the organic material on the bottom with sediments for additional millions of years (Fig. 2.3). This creates the heavy layers of sediments and eventually rocks that cover the oil and insure that the organic material will be pressure-cooked for 100 million years or so. The result is that the long complex molecules of organic matter are "cracked" into shorter molecules.

Clearly these conditions did not occur very often, but there is still another important consideration. Because oil and gas are considerably lighter than the sediments that overlay them, they tend to migrate upwards from their "source rocks" through porous sediments (Fig. 2.4). It is only where they encounter impermeable "caps," such as special types of sandstone or salt, that they are captured by a "trap rock," forming the reservoirs that we can exploit. Thus, these rare and special series of occurrences that happened millions of years ago are essential to modern industrial life as we know it, and yet they were so geologically rare as to almost not have happened! This is why oil is relatively rare today and why we cannot count on finding very much more!

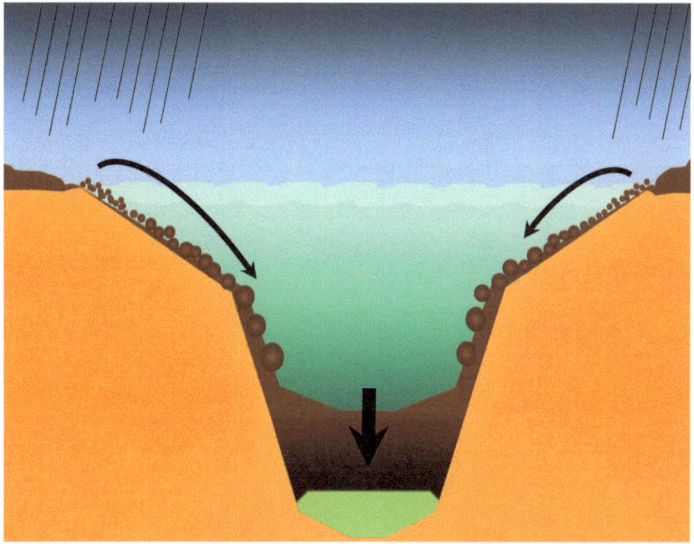

Fig. 2.3 Deposition of geological sediments increases pressure and temperature, providing the conditions for the decomposition and transformation of organic matter into hydrocarbon minerals (source rocks); modified and published with permission of Colin Campbell

2.3   The Evolution of *Homo sapiens*

Our story of the development of the hominid line leading to humans starts some 4 million years ago, in Africa. The same as other adventurous species, hominids developed adaptations to their changing environment. They passed through several species to arrive at a rather primitive version of modern humans about two and a half million years ago, when *Homo habilis* was making use of crude stone implements. Only about 500,000 years ago, large brained descendants evolved most likely from *Homo heidelbergensis*, and 300,000 years later, two species had appeared: the *Homo neanderthalensis* and the early *H. sapiens* (Smithsonian Institution 2010); their many physical and cultural common traits still raise the question whether they should be considered as two subspecies of a single species. The early sapiens lived in Africa until about 60,000 years ago (Genographic Project 2011), then began to spread out into Asia and Europe, leaving traces of their existence in places such as the rock shelters of Crô-Magnon, in France, or the Qafzeh and Skhul caves in Israel (Harder 2002). Little is known about the interactions between the early sapiens and the Neanderthals; some theories say that the former probably indulged in a degree of genocide, ridding the world of their hominid cousins, while other theories have claimed that both subspecies actually interbred into modern humans. Recent DNA tests have supported the latter argument to some degree (Than 2010).

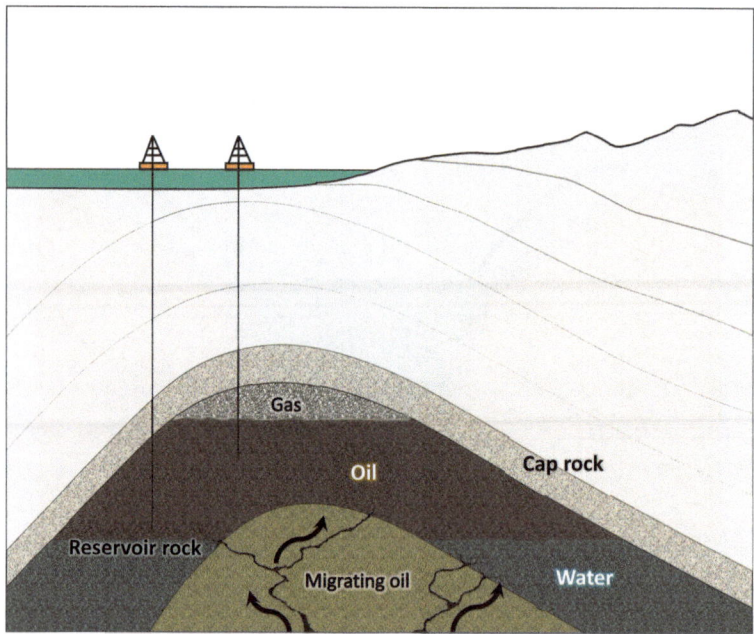

Fig. 2.4 The resulting hydrocarbons are lighter than water and tend to migrate upwards through cracks and pores in the lower sediments until they find an impermeable rock (cap rock) that prevents them from escaping to the surface. Additionally, the layer embedding the hydrocarbons has to be sufficiently porous to allow their "free" movement (reservoir rock); otherwise the extraction would require the injection of gas at very high pressures, becoming costly or unprofitable; modified and published with permission of Colin Campbell

2.3.1 Civilized Humans

Like their ancestors, the early sapiens started as nomadic hunters, fishermen, gatherers, and cave dwellers. But imitation, learning, and teaching, a trait that many other species do not have, resulted in "cultural evolution," which would allow the development of very different behavioral paths, such as sedentism and plant domestication. These practices provided the basis for the emergence of agriculture, a new form of livelihood that appeared about 10,000 years ago in the Fertile Crescent (Mann 2011). Agriculture gave rise to early political structures that were increasingly able to control the future of static communities. This more sedentary existence may have influenced the later development of metallurgy and mining operations when the farmers started looking for materials to manufacture better tools. About seven thousand years ago, different cultures learned how to smelt gold and copper, perhaps largely for ornaments (Radivojević et al. 2010). Then a metallurgical breakthrough came into effect, only three and a half thousand years ago, when an alloy of copper and tin made possible the manufacture of tougher bronze weapons and tools. Thus the so-called Bronze Age started, marking the rise of civilizations. If we were to view the fossil record of life on Earth since the Cambrian, when limpets were born, as a single day, civilized humans appeared only at twenty seconds to midnight— a very recent arrival by all means.

Despite these and other quite remarkable adaptations, these early humans still had to deal with the cataclysmic fluctuations of the environment that occurred from time to time. In some cultures, religion may be regarded as an early attempt to fulfill this objective, offering the possibility of explanation and control. In many civilizations, agricultural activity is strongly tied to religious practice, organizing the time for sowing and harvesting, as well as providing the social rules for the storage and distribution of crops and other foods (Mann 2011). Moreover, currency and accounting systems have been directly related with grain storage. As the use of metals became widespread, another development saw the use of gold and silver as means of exchange. Metal objects, especially ornaments, were highly appreciated in many cultures. Gradually metals became the medium of exchange in Greece, India, and China, among other civilizations (Schaps 2004), its value being set by its natural scarcity. By the Middle Age, Venice had evolved into an important trading town, importing goods from the eastern Mediterranean. In the sixteenth century, the Spanish Crown had embarked itself on the conquest of the Americas, where large deposits of gold and silver existed.

However, the most important process resulting from the metallurgical breakthroughs that started during the "Bronze Age" was not the initiation of the quest for gold around the globe by our curious hominid but to begin the path towards industrialization through the exploitation of metals. Other metallurgical breakthroughs occurred after charcoal and bellows were used to reach sufficiently high temperatures to smelt iron. This metal could be worked into vastly stronger weapons and implements, followed later by a still stronger material, steel. In Europe, the "Iron Age" followed the "Bronze Age" until a civilization built around steel began to flourish only about 300 years ago, less than one second to midnight in our geologic clock.

2.4 Industrialization: The Rise of *Homo hydrocarbonum*

The last chapter of our story opened only 200 years ago—on the stroke of our geological midnight—having its origins in Britain, where energy from millstreams was harnessed by the waterwheel to drive looms for weaving cloth. At first, firewood was the fuel for smelting the metal, which in certain countries, such as Denmark, led to damaging deforestation. But later, a new fuel was found in the form of coal, lumps of which, known as sea-coal, were collected from beaches before it was mined in shallow pits. Sea-coal had been known for a long time, but the increased demand for fuel led to mining coal in shallow pits. The pressure to deepen the mines below the water table led to the development of steam-driven pumps to drain mine workings. The pump was adapted by feeding steam into the cylinders, causing the pistons to turn a wheel. The steam engine resulted, which was soon adapted to drive a locomotive ushering in the age of rail, which expanded trade and travel greatly.

Sail gave way to steam, opening up world trade. Inventors in continental Europe were looking for an efficient way to insert the fuel directly into the cylinder. These developments led to the invention of the internal combustion engine at the beginning of the nineteenth century. In the 1870s, inspired by the success of the engines designed by the Belgian inventor Etienne Lenoir, a German engineer by the name of Nicolaus Otto built the world's first four-stroke engine. At first, it relied on "illuminating gas" distilled from coal before turning to gasoline refined from crude oil. Using the Otto cycle, Carl Benz built a road vehicle completely propelled by gasoline in the next decade (Eckermann 2001).

In short, the advent of engines, together with the ongoing social struggles of the epoch, led to a new form of economic organization, now known as capitalism, as the mill-owners accumulated wealth by the use of machinery that cost less than human energy. The burst of new capital stimulated expansion and the search for new markets, which in turn prompted the growth of empires, notably those of Britain, France, and Russia. Coal in England and Germany was abundant and relatively easily mined, so the economic and the energy growth fed each other. In parallel with that, a great increase in the use of debt and credit came into effect, requiring a confidence in the system and its growth and progress, sometimes underpinned by military force. It in turn brought an expansion of usury that fed new money into the system, fueling further economic growth. More fuel could always be found to do the physical work required to allow business expansion, profits, and the repayment of earlier debts.

The Industrial Revolution brought its own pressure for migration from Europe to the New World. Some adventurers may have gone enthusiastically in the quest of a new life of opportunity, but most were driven out in abject desperation. This desperation was brought about partly by changed social circumstances of economic disparity that arose directly from population growth and indirectly from the new capital and technology provided by fossil energy. Think of poor Ireland, the homeland of Colin Campbell, where a combination of adverse land tenure, disparate wealth, displacement of labor by machines, and a potato disease brought wholesale emigration such that the population is now half what it was 150 years ago. Ireland pays great respect to its dead, holding celebrations, known as wakes, to mark funerals. The departing emigrants were treated to what was known as an American wake, as grieving families lost their offspring forever. It was a case of desperation, not adventure. Even though the new energy would also transform agriculture, providing the food for a growing population, Europe lacked the land to support its own people with the farming methods then available that relied on natural nutrients. This is underlined by the fact that it was economically viable for Europe to import soil nutrients, derived from the excrement of seabirds in Chile and Peru, in sailing ships. Indeed a pigeon in France was valuable more for its droppings than its flesh. Synthetic fertilizers came later when Norway harnessed hydroelectric power to extract nitrogen from the air, which was then vastly improved by the petroleum-based Haber Bosch process developed in Germany before and during the First World War. Mechanized farming, combined with irrigation, itself largely relying on oil-driven pumps, led to a huge increase in food supply—but required fewer farmers. Newly engineered plant types gave increased yields, but had a voracious appetite for synthetic nutrients and water.

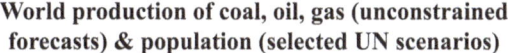

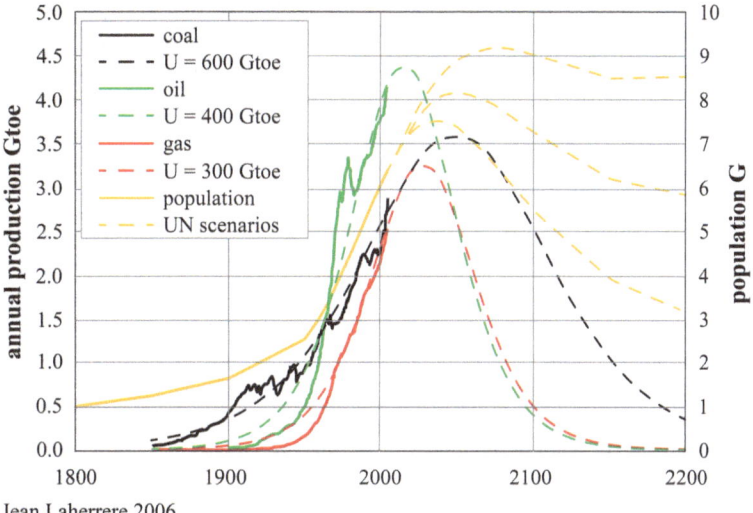

World production of coal, oil, gas (unconstrained forecasts) & population (selected UN scenarios)

Jean Laherrere 2006

Fig. 2.5 Population growth and production of fossil fuels. The production of fossil hydrocarbons played a large role in the emergence of the industrial system which supports a large part of the seven billion people in the world

Nevertheless, at the beginning of nineteenth century, the world was still a large place. The world's population at the time of Christ was about 300 million, and it had no more than doubled until the nineteenth century, standing steady below 500 million, with a slight dip in fourteenth century from the Black Death. Most people lived simple, sustainable rural lives on whatever their particular region could support. Their energy came mainly from their own muscles, although supplemented by that from slaves, draught animals, wood, wind, and water power. But then population doubled during the first half of the nineteenth century, as coal-based industrialization brought economic expansion. The arrival of oil in the second half of the century had a more dramatic effect; it made possible for the population to expand sixfold, exactly in parallel with the growth of oil in only 150 years (Fig. 2.5).

Two world wars and other partly related conflicts did something to cull the population: with 8 million casualties in the First War, 41 million in the Second War, and tens of millions in genocides and political exterminations around the world, the loss had no perceptible impact on the overall population trend. The expansion of fossil-fueled machines was so aggressive as if everyone had an unfed and barely paid team of slaves to do his manual work for him or her. Modern sapiens, having tapped the world's fossil energy supplies of coal, oil, and later natural gas, became immensely successful by any biologist's criteria. Of course it has been a success at the expense of other species, which are being wiped out by human destruction of essential environments at rates equaling those in the geological record when massive volcanic eruptions blotted out the sun for centuries or asteroid impacts shook the planet.

It is worth noting that the growth in population was achieved more by rising longevity than increased fertility. Average life expectancy in 1950 was around 48 years, but by 2010, it had risen to 68 years. It was highest in the wealthy nations, but was offset there by declining birth rates, consequent upon the emancipation of women, many of whom preferred paid employment to raising families at home. The fertility rate in much of the developed world is now running at 1.71 children per woman, far below the replacement rate of 2.1 children per woman (UN DESA Population Division 2011). As a consequence, within a generation or so, few of the indigenous people in such countries will have brothers, sisters, cousins, aunts, or uncles. The accumulated wealth of past generations will flow to the survivors, or to be taken by the state in inheritance tax. Furthermore, the industrial populations are aging, placing an ever-heavier burden on the health services and indirectly the young work force that supports them. The aged have become a major voting power, naturally pressing for improvements to their lot.

The energy-rich societies need new immigrants to support them, so they welcome cheap labor inside their countries, despite the ethnic tensions it brings. Many workers migrate from low energy countries in the belief that they will be absorbed eventually as the economy expands ever onward to accommodate them. There is no shortage of supply coming from the depressed urban centers of China, East Europe, Africa, Latin America, and the Indian subcontinent. Every night, flimsy crafts try to cross the Straits of Gibraltar to deliver illegal immigrants onto the Spanish beaches. By no means do all make it. Frozen would-be immigrants have even been found in the undercarriage cowlings of aircraft. There has been no letup on the Rio Grande until quite recently. On Christmas Day 2001, 500 illegal immigrants attempted to storm the Channel Tunnel in their bid to reach England. A new gruesome trade of human smuggling has been added to the global market.

However, as the resource constraints, especially of cheap oil, lead to long-term recession since 2008, there is less and less room to accommodate both the surviving indigenous populations and the new immigrants with their descendants. This curious outcome is seemingly consistent with the economic principles of "supply and demand" and arguments about "efficient allocation" of scarce resources: in this case labor had been flowing from places where it is "overabundant" and cheap to places where it is less abundant and more expensive. Now, there are few opportunities anywhere for more and more of the world's poor, unless the global system of wealth distribution undergoes meaningful changes.

Falling fertility is not being experienced in the Middle East, however, where a combination of unearned oil revenue and traditional family patterns is causing a population explosion. The resulting youthful generation faces a difficult future, depending on the uncertain disbursement of the proceeds of oil revenues by essentially feudal governments, over which the average people have no control. As the population expands, the share of the patrimony has to decline, as indeed will the patrimony itself as oil depletion grips even these countries in the years ahead. There is not much gainful employment to be had in the barren deserts.

The new energy freed people from drudgery in the energy-rich countries (including those like the Japanese who learned how to import and use it), helping to make possible great achievements in science, medicine, literature, and general culture. The amazing pace of progress made a deep impression, leading people to believe that humans were the masters of their environment in a world of near limitless resources to be bent to their will. These notions were enshrined in the new subject of economics, which enunciated supposedly immutable laws of supply and demand. The world was seen as a marketplace such that if the price of wheat should rise, the farmer would grow more in the next sowing, returning the system to an everlasting equilibrium. Financial management became more sophisticated, although not always with positive results. Economic growth has led to enormous new wealth, however unevenly distributed. Religious teachings further consecrated the special position of humans in the universe, being perceived to be closer to God than the birds and the beasts of the field. This, together with the emergence of subjects such as eugenics, led to notions that certain races were better than others, and that wars and genocide were an essential part of evolution.

The United States emerged supreme from the Second World War, when a weakened Britain and France voluntarily surrendered their empires. The new economic empire, built with, by, and for the dollar, replaced the old European empires, spreading its unseen financial tentacles throughout the world. There was a great disparity, not only between the rich and poor nations of the world, but within the rich countries themselves. Manufacturing was progressively transferred to the poorer countries so as to benefit from what almost amounted to slave labor, while the profits, partly in the form of mysterious credit, flowed home to the wealthy nations, which became ultra-consumers. Hairdressers in affluent capitals served their clients, both arriving in large automobiles. Property values soared as neo-palaces were built for the new executive kleptocracy and as hitherto humble people developed new aspirations. Conditions in many poor countries deteriorated, especially in the growing urban agglomerations. Political tensions arose in many places. Colombia is not the only country to face civil war as a consequence of globalism, caused in its case by a trade in narcotics made from its traditional cultivation of cannabis and coca species. Mexico, considered an economic miracle during the 1940s, 1950s, and 1960s, is now living a similar situation: drug cartels and other criminal organizations have attained enough economic and military power to disrupt the life of the entire country. Moreover, drugs neither are the only commodities which have been involved in international conflicts nor are they the only products that are routinely smuggled overseas into both the developing and the developed world: oil has been widely linked to armed violence and illegal traffic too (Margonelli 2007, 2009; Alic 2012; CBS News 2009).

Oil was followed in turn by gas, increasingly used for electricity generation, which brought power and light to households throughout the world. It opened the door to world electronic communication, and eventually the abuse thereof through television, which helped condition the modern consumerist mind-set. Industrialization made formerly self-sufficient peasants into landless wage earners, consumers, and taxpayers, leading in turn to the growth of capital, debt, and usury, which dominate the modern world, giving birth to an increasingly globalized virtual economy. Globalization is changing the world, but a globalized world is possible only if the massive infrastructure of the transportation and communication systems is in place, and this infrastructure is fed through vast quantities of cheap oil that are themselves transported through big container ships every day. The end of cheap oil will disturb globalization certainly.

Now, as the twenty-first century dawns, we face the onset of the natural decline of the premier fuels that made all this possible, and we do so without sight of a substitute energy that comes close to matching the utility, convenience, and low energy and hence monetary cost oil and gas. Today, 32 billion barrels (Gb) of oil a year support 7 billion people with an energy supply equivalent to that of billions of slaves working around the clock. From the humble beginnings of hominid hunters, this new subspecies, the *H. hydrocarbonum*, evolved to become the sole surviving subspecies of the sapiens, as everyone from the Borneo native to the Manhattan commuter relies on petroleum to varying degrees. So far, there is little sign of a successor. It remains to be seen if we will be the only species in over 500 million years of recorded history to evolve backwards from complexity to simplicity.

We are not about to run out of oil, as we have about as much oil left as we have used so far. But we are entering the second half of the age of oil, a time when the amount available, rather than increasing every year, as has been the case for 150 years, will instead be decreasing. So there is a little time in which to adjust to declining supplies. Our challenge is to cut demand to match or, better, fall below the depletion rate and to use what we have left to ease the transition to whatever follows. The first step in that direction is to determine what the depletion rate is and to better inform ourselves about the resources with which nature has endowed us.

References

Alic J (2012) Greece: oil smuggling helps define the "parliamentary mafiocracy." http://oilprice.com/Energy/Energy-General/Greece-Oil-Smuggling-Helps-Define-the-Parliamentary-Mafiocracy.html. Accessed 30 May 2012

CBS News (2009) Feds bust Mexico-U.S. oil smuggling scheme. In: CBS News. http://www.cbsnews.com/2102-201_162-5231681.html?tag=contentMain;contentBody. Accessed 30 May 2012

Eckermann E (2001) World history of the automobile. SAE International, Warrendale

Genographic Project (2011) Atlas of the human journey. In: The Genographic Project. https://genographic.nationalgeographic.com/genographic/lan/en/atlas.html. Accessed 18 May 2012

Harder B (2002) Did humans and Neanderthals battle for control of the Middle East? In: National Geographic News. http://news.nationalgeographic.com/news/2002/03/0305_0307_neandertal.html. Accessed 29 May 2012

Klemme HD, Ulmishek GF (1991) Effective petroleum source rocks of the world; stratigraphic distribution and controlling depositional factors. AAPG Bull 75:1809–1851

Mann CC (2011) Göbekli Tepe. In: Pictures, more from national geographic magazine. http://ngm.nationalgeographic.com/2011/06/gobekli-tepe/mann-text/1. Accessed 2 Aug 2012

Margonelli L (2007) Blood oil. http://pipeline.blogs.nytimes.com/2007/01/08/blood-oil/. Accessed 30 May 2012

Margonelli L (2009) Oil smuggling: is it time to start worrying yet? In: The Atlantic. http://www.theatlantic.com/technology/archive/2009/08/oil-smuggling-is-it-time-to-start-worrying-yet/23303/. Accessed 30 May 2012

Radivojevi M, Rehren T, Pernicka E et al (2010) On the origins of extractive metallurgy: new evidence from Europe. J Archaeol Sci 37:2775–2787. doi:10.1016/j.jas.2010.06.012

Schaps DM (2004) The invention of coinage and the monetization of ancient Greece. University of Michigan Press, Ann Arbor

Smithsonian Institution (2010) Human evolution timeline interactive. In: The Smithsonian Institution's Human Origins Program. http://humanorigins.si.edu/evidence/human-evolution-timeline-interactive. Accessed 30 May 2012

Than K (2010) Neanderthals, humans interbred—first solid DNA evidence. In: National Geographic News.http://news.nationalgeographic.com/news/2010/05/100506-science-neanderthals-humans-mated-interbred-dna-gene/. Accessed 29 May 2012

UN DESA Population Division (2011) World population prospects, the 2010 revision. In: United Nations Department of Economic and Social Affairs, Population Division. http://esa.un.org/unpd/wpp/index.htm. Accessed 30 May 2012

Bibliography

This chapter is primarily based on the following sources:

Campbell CJ (2010) The post peak world. http://iprd.org.uk/wp-content/plugins/downloads-manager/upload/The%20Post-Peak%20World.pdf

Campbell CJ (2003) Essence of oil and gas depletion. Part 5. Multi-Science Publishing Co. Ltd., Brentwood

Population and political issues are based on

Campbell CJ (2002a) Petroleum and people. Popul Environ 24:193–207. doi:10.1023/A:1020752205672

Campbell CJ (2002b) The assessment and importance of oil depletion. Energ Explor Exploit 20:407–435

Chapter 3
The Nature of Petroleum

In the previous chapter, we have depicted the evolutionary processes of oil formation and the industrial civilization. Of course, many other minerals have been formed through geological processes, but fossil fuels, and oil in particular, have some special characteristics. In this chapter we will explore the physical and chemical properties, resulting from the geological processes previously illustrated, which make fossil fuels unique.

As we have seen previously, oil and gas are fossil remains—in all the sense of the word—of prehistoric organisms that proliferated during the Paleozoic, Mesozoic, and Cenozoic periods. Then, it is no surprise that oil and gas have inherited many chemical and physical characteristics from the living creatures that participated in their formation. Petroleum comes from porous reservoirs (sand or carbonate rocks), hence its name, which means "rock oil."

Even though petroleum is a mineral, it has an organic origin. As those on diets soon come to find out, every living organism is largely made up of proteins, carbohydrates, and lipids (fats), all of which are based on carbon, hydrogen, oxygen, nitrogen, and sulfur. Back in the beginning of the nineteenth century, chemists began to study these compounds in order to provide science with the chemical basis for the understanding of life, giving birth to the branch of chemistry called organic chemistry. They soon realized that the overwhelming majority of compounds occurring in organisms are compounds of carbon and that the study of carbon by itself offered a broad domain of knowledge entirely independent of biology. Thus, the subject of organic chemistry came to be redefined simply as the chemistry of carbon compounds (Meinwald 2003).

3.1 Carbon: The Basis of Life

Indeed, carbon has some very remarkable properties that are worth exploring in some detail. Found in abundance in the sun, stars, comets, and atmospheres of most known planets, it is the fourth most abundant element in the universe by

C.A.S. Hall and C.A. Ramírez-Pascualli, *The First Half of the Age of Oil: An Exploration of the Work of Colin Campbell and Jean Laherrère*, SpringerBriefs in Energy, DOI 10.1007/978-1-4614-6064-0_3, © Springer Science+Business Media New York 2013

mass after hydrogen, helium, and oxygen and the 15th most abundant in the Earth's crust (Croswell 1996). It is present in all known organisms, being the second most abundant element by mass in the human body (about 18%), surpassed only by oxygen (Chang 2007). Despite its notable abundance, what really makes carbon unique is its outstanding ability to be combined in different ways; carbon atoms have six electrons, four of which are available to form covalent chemical bonds or to use some chemical jargon, carbon is a tetravalent element. Due to this configuration, carbon is able to form more different and more complex compounds than any other element: the Chemical Abstracts Service (CAS) had registered more than twenty-four million organic compounds by 2007 (Lipkus et al. 2008), and this quantity appears to be a minuscule fraction of the theoretically possible number under standard conditions (Ertl 2002). Carbon can form small, large, repetitively simple, or highly complex molecules at the temperatures commonly encountered on Earth. Indeed, this characteristic has given place to the development of an entire protocol to name carbon compounds; this systematic nomenclature is responsible for names such as cyclohexanecarbaldehyde or 3-methyl-propyloctane-4, to cite only two conservative examples. To complicate things further, carbon has three naturally occurring isotopes: the stable versions ^{12}C and ^{13}C and the radioactive variety ^{14}C, decaying with a half-life of about 5,730 years (Godwin 1962).

The same propensity to establish bonds causes carbon compounds to be stable. Relatively large amounts of energy are required to make or to break a carbon bond. The consequence is that organic compounds do not degrade fast under standard conditions, and organic tissues do not dissolve in water or burn in the air at average atmospheric temperatures. However, all carbon compounds are susceptible to degradation by oxidation (combining with oxygen), but these reactions occur very slowly at normal conditions—as in the respiration process in plants and animals. In some sense, we can say this behavior regulates the entire process of life itself. Beyond normal conditions or in the presence of an ignition source, carbon compounds lose their stability and become combustible: they release heat to the surroundings at a fast rate. All carbon compounds are combustible in the air at the right temperature (as low as 185°C or 365°F for acetaldehyde). Conversely, carbon compounds would burn at standard conditions in the presence of enough oxygen (e.g., diamonds burn when immersed in liquid oxygen).

Probably, carbon is the first nonmetallic element that had some useful applications for the human species: it certainly played an important role in the early metallurgy that allowed the start of the Bronze Age. Carbon has been known since antiquity. The Egyptians and Sumerians, 5,000 years ago, used carbon as charcoal for the reduction of copper, zinc, and tin ores in the manufacture of bronze, as well as domestic smokeless fuel. Charred wood was used later with medicinal and aseptic purposes: Hippocrates and Pliny record the use of charcoal to treat a wide range of complaints including epilepsy, chlorosis, and anthrax. Its name is derived from the Latin *carbo*, meaning charcoal or coal (Carbon Materials Group 2012).

3.2 Hydrocarbons: A Very Special Mix

Oil and natural gas, together with other related substances such as coal, asphalt, and bitumen, are hydrocarbons, which means that they are formed basically by hydrogen and carbon combined in various ways. Actually, the majority of hydrocarbons that occur in nature are present in crude oil, so there is no such a thing as a chemical formula for it. This lack of standardization introduces uncertainty in the quantification of oil and of the energy that we can obtain from it. As we will see time and again, when it comes to definitions about oil, uncertainty is the name of the game.

Due to its carbonaceous nature, the members of this remarkable family of substances are relatively stable but exhibit many combinatorial possibilities. They can dissolve intimately in each other, depending partly on the ambient temperature and pressure, and mix with oxygen to form CO_2: they are, so to speak, born, transformed, degraded, and disappear, having in this respect a sort of life cycle that mirrors their organic origins. A very important part of this "life cycle" is the interaction between hydrocarbons and oxygen, an element that is indeed abundant in the atmosphere. Hydrocarbons react chemically with oxygen at temperatures above 200°C (329°C), a combustion process where carbon and hydrogen combine with oxygen to produce CO_2, water, and heat majorly. Since both hydrogen and carbon are combustible—that is, they react with oxygen during a reaction that liberates heat—hydrocarbons are able to produce a very large amount of heat when burning.

Hydrogen, the other component of hydrocarbons, is the only element that is able to challenge carbon in terms of the number of compounds where it appears. Hydrogen not only appears in most organic compounds but also is able to bond with all the elements that occur in nature, an ability that carbon does not have. Moreover, hydrogen bonds are likely to have greater energy values than carbon bonds. Hence, the more hydrogen bonds present in the molecules, the more heat will be released. This fact has led many people to believe that our civilization can be fueled by molecular hydrogen (H_2). However, even though H_2 is combustible, hydrogen atoms are not available in nature: you have to separate them from other compounds through a process that requires a lot of energy. Hence, hydrogen is not a source of net energy at all. Furthermore, hydrogen is very hard to enclose and leakages are dangerous and difficult to detect, as the Hindenburg accident showed to the world in 1937.

Methane is the hydrocarbon with the greatest proportion of hydrogen relative to carbon (4 to 1). This means that methane has the highest calorific value per unit of mass among hydrocarbons. However, methane is a very light gas, so a kilogram of methane occupies a larger volume than a kilogram of heavier hydrocarbon gases, such as propane, or of other liquid hydrocarbons, such as kerosene or gasoline. Due to its low density, methane is usually not used to fuel engine vehicles: it takes a very large tank to hold significant quantities of it.

Given that carbon is combustible, many organic compounds are combustible too. However, their calorific value is less than that of hydrocarbons because (1) their hydrogen content is lower and (2) most of them contain oxygen, which is a comburant element but not a combustible substance. Carbohydrates are a good example of organic compounds that we can find easily in the environment. They are present in living tissues, in most of our foods or in wood, a material that humans have used as an energy source for millennia. However, the presence of oxygen in carbohydrates means that these compounds have a lower energy density than hydrocarbons do. The same applies to alcohols, largely known to humans in different cultures as the product of fermentation of glucose, such as the ethanol found in alcoholic beverages. In these substances, the presence of the hydroxyl group—that is, an atom of oxygen connected to an atom of hydrogen (OH)—affects the energy content per unit of mass negatively. Thus methanol and ethanol have a lower energy density than methane, ethane, or gasoline. Therefore, burning biomass or biofuels, no matter how ingenious mechanisms we invent, is not as sophisticated as many people have been led to believe. In chemical terms, both "solutions" are not a step "forward," but a complex and elaborated way to go back.

3.2.1   Classification of Hydrocarbons

Since hydrocarbons are very important compounds for our economic system and our lifestyle, before we continue our exposition, we think it is useful to provide a general classification of organic compounds. This way, you will have a wider perspective on what the terms "oil" and "gas" mean in this context. The International Union of Pure and Applied Chemistry (IUPAC) has classified hydrocarbons into four groups according to their structure (IUPAC 1993):

1. *Saturated hydrocarbons.* The simplest form of hydrocarbons, they form a linear chain of carbon atoms with single bonds between them; "saturated" means that there is no place for extra carbon atoms in the linear chain, that is, the compound is saturated with carbon. They are also called "alkanes."
2. *Unsaturated hydrocarbons.* Like their saturated cousins, these molecules also form linear chains but with double or triple bonds between carbon atoms; one of the double or triple bonds can hold another carbon atom in the structure, so the chain is not completely saturated. Double-bonded compounds are called "alkenes" (olefins is an old name for them) and triple-bonded compounds, "alkynes."
3. *Cycloalkanes.* Instead of forming simple chains, carbon atoms form a ring or cycle with single bonds between them.
4. *Aromatic hydrocarbons.* In these configurations, at least one bond in the ring of carbon atoms is something between a single and a double bond that we will call here a "hybrid" bond. The name "aromatic" comes from the solubility in air that these compounds exhibit, producing a strong smell. Turpentine, a resin extracted from some pine trees, is a familiar example.

However, due to tradition, a different classification is used in the oil industry. Table 3.1 links both classifications, so you will not be overwhelmed by chemists or oilers.

Hydrocarbons can also be classified according to the number of carbon atoms that they have. Molecules with only one atom of carbon are denoted by the prefix "meth-";

Table 3.1 Hydrocarbon classification: common name and IUPAC denomination

Common name	IUPAC denomination	Structure
Paraffins	Saturated hydrocarbons	Linear saturated
Naphthenes	Cycloalkanes	Cyclic saturated
Aromatics	Aromatic hydrocarbons	Cyclic with "hybrid" bond
Asphaltenes	Precipitate with no formal definition	No formal structure

Table 3.2 IUPAC classification of some simple hydrocarbons

Carbon atoms	Alkane	Alkene	Alkyne	Cycloalkane
1	Methane	–	–	–
2	Ethane	Ethene (ethylene)	Ethyne (acetylene)	–
3	Propane	Propene (propylene)	Propyne (methylacetylene)	Cyclopropane
4	Butane	Butene (butylene)	Butyne	Cyclobutane
5	Pentane	Pentene	Pentyne	Cyclopentane
6	Hexane	Hexene	Hexyne	Cyclohexane
...	...	...	...	...
10	Decane	Decene	Decyne	Cyclodecane

Common names in parenthesis

if two atoms of carbon are present in the molecule, the compound is denoted with the prefix "eth-"; three atoms of carbon, corresponds to "prop-"; four atoms, "but-"; five, "pent-"; six, "hexa-"; and so on. If the compound is linear and saturated (single bonds) it receives the suffix "-ane"; if it is unsaturated and has a double bond, the suffix is "-ene"; but if it has a triple bond, "-yne". If it forms a cyclic structure, it receives yet another prefix: "cyclo".

In Table 3.2 we depicted only the simple cases where there is only one double or triple bond in the molecules. Things grow more complicated if there are more than one double or triple bond, when there are ramifications instead of simple carbon chains, or when aromatic rings appear. In general, all these compounds occur or can be synthetized from petroleum. It is no surprise that oil is a basic input for many of our industries.

3.3 The Components of Petroleum

Let us focus now on the major compounds in petroleum. Here we will use the industrial classification—instead of the IUPAC classification—because that is what you will find most likely when reading news or reports about oil and natural gas.

3.3.1 Natural Gas

Natural gas is formed mostly of the simplest hydrocarbon, methane, which is relatively soluble in water. In methane each of the four electrons of the carbon atom is bonded with a hydrogen atom, so its chemical formula is simply CH_4; when natural gas contains methane almost exclusively, it is called dry gas, but if it contains liquid hydrocarbons, such as propane and butane, it is called wet gas. The isotopic composition of methane—that is, the proportion between the different carbon isotopes—generally reflects the degree to which it has been subjected to rising temperature and pressure on burial. Some natural gas deposits contain hydrogen sulfide—the mix is then known as sour gas, or nitrogen and carbon dioxide, depending on depositional conditions and the effects of alteration. Small amounts of helium and argon are also sometimes present. Natural gas is normally found in close association with an oil accumulation. It is highly compressible, such that its volume may be reduced by a factor of 200–300, which incidentally means that it can provide a valuable drive mechanism to expel associated oil from a reservoir.

Hydrate seeps are special deposits of methane in icelike solid conditions. These seeps are found in the seafloor of the Arctic and in deep oceanic environments. Some hopes have been entertained for exploiting such deposits, which may be very large, but they are unlikely to be fulfilled because methane molecules become trapped into a solid matrix of ice and have no opportunity to continue its migration and accumulation in commercial quantities (see Fig. 2.4). Therefore, the concentration of methane is likely to be too low to be economically exploitable.

3.3.2 Paraffins

Paraffins are saturated hydrocarbons (or alkanes), which are quantitatively the most important, making up 50–60% of most oils. They form a linear molecular chain with the general formula C_nH_{2n+2} and occur in three states: gas (1–4 carbon atoms), liquid (5–15), and solid (above 15).

So-called n-paraffins, with odd numbers of carbon atoms, are synthesized primarily in living organisms; such molecules found in oil are true biological markers inherited from the living organisms from which they were derived. C_{15}, C_{17}, and C_{19} characterize microscopic organisms including algae, whereas molecules above C_{21} typify plants. These chemical links give the game away, showing that oils come primarily from algae. Another group of molecules, the iso-alkanes with a branched structure, include pristane (C_{19}) and phytane (C_{70}), are derived from chlorophyll in living organic material. They also demonstrate the link.

3.3.3 Aromatic Hydrocarbons, Naphthenes, Resins, and Asphaltenes

Aromatic hydrocarbons have a ringed molecular structure with a "hybrid" bond (see Sect. 3.2.1). They are so named because of their pleasant smell. Benzene (C_6H_6) is the simplest aromatic hydrocarbon. The naphthene family also has a ringed structure but only with single bonds between carbons. This family is commonly associated with sulfur compounds, giving them the exceedingly unpleasant smell of bad eggs. Resins and asphaltenes are complex compounds with high molecular weight, rich in nitrogen, oxygen, sulfur, nickel, and vanadium. They are mainly the products of the chemical alteration of ordinary oils.

3.4 The Physical Properties of Petroleum

As we have seen before, there is no chemical formula for petroleum, so it should be no surprise that its physical properties vary widely. Contrary to the popular idea, petroleum is not a black liquid, at least not always. Petroleum can occur as a solid, liquid, or gas, depending on the ambient temperature, pressure, and chemical composition; each phase may contain dissolved elements of the others. Oil is a liquid hydrocarbon and generally contains fractions of the gaseous and solid phases in solution; on the other hand, natural gas sometimes contains dissolved liquids. The different phases may separate naturally and can be extracted by processing. In addition, the color of crude oil varies greatly depending on its composition. It is usually black or dark brown, although it may be yellowish or even greenish.

3.4.1 Density

Density mainly reflects the chemical composition of the oil. In the Anglo-Saxon world, density is traditionally measured under a scale set by the American Petroleum Institute (API) that measures how heavy or light a petroleum liquid is compared to water (API gravity), with most oils being in the range 15–45° API (0.9–0.7 specific gravity); if an oil is less than 10° API, it will sink in water, and it will float if greater than 10° API. The heavier oils are rich in resins, asphaltenes, and sulfur, whereas the lighter oils tend to contain dissolved gas. Heavy oils are generally dark brown or green in color, whereas the light oils may be almost as clear as refined gasoline.

As a term, "heavy oil" is applied to oils with a density below 10° to some 28° API. There is unfortunately no standard industry definition of the density threshold for heavy oil, which is a cause of much confusion. The sulfur content may be as high as ten percent. Heavy oils have various origins but most commonly are normal oils from which the light fractions have been removed by water leaching, oxidation, or microbial degradation. Huge deposits of heavy oil and bitumen occur in Eastern Venezuela, Western Canada, and Siberia, forming important lower-quality resources that we have been in the need to exploit recently.

3.4.2 Viscosity

The inverse of fluidity, viscosity generally increases with density and decreases with the dissolved gas content and a higher temperature which is the main factor. Viscosity is measured in centipoise and ranges from 1 cP to more than 10,000 cP. This property is related to the pour point (the lowest temperature at which a liquid will pour or flow under certain conditions), which is linked to the paraffin content. The pour point is important because it provides a rough estimate of the lowest temperature at which oil is readily pumpable. Some crudes, especially waxy ones of non-marine origin, become pasty and solid below about 10°C (50°F). Oils from Athabasca, Canada, and the Orinoco basin, in Venezuela, have about the same API gravity (<10° API). In Athabasca, where the reservoir temperature remains between 5°C and 10°C, the oil has a large viscosity (1,000,000 cP) and does not flow, but in the Orinoco basin, the reservoir temperature is much higher (55°C) and the oil flows due to its small viscosity (2,000 cP). Alaskan oil is another interesting example; the heat generated by a high throughput is needed to prevent the oil in the Alaskan pipeline from solidifying.

3.4.3   Solubility

A third important property is solubility, namely, the ability of the several fractions to mutually dissolve in each other. In particular, large amounts of gas can be dissolved in oil; the solubility of methane in water varies considerably with pressure. A measure of the gas content dissolved in oil is the gas–oil ratio (GOR), which may be as high as 6,000 cubic feet per barrel (1,000 m^3/t). The ratio varies inversely with density and rising pressure. Where conditions approach the bubble point, the gas separates to form a gas cap above the oil accumulation. The dissolved gas increases the volume of the liquid, and a so-called formation volume factor has to be applied to convert volumes of oil in the reservoir to those at the surface where the gas comes out of solution.

Solubility measures and volume estimates are becoming more and more important topics as natural gas becomes a more valuable resource. For example, the geopressured brines of the Gulf Coast have methane resources estimated at 50,000 trillions cubic feet (Tcf). In 1977, Dr. Vincent E. McKelvey, who was then director of the US Geological Survey, said that volume was "as much as 3 000 to 4 000 times" the amount of natural gas the United States would consume in that year.

As we are noticing, petroleum is a slippery substance in more than one sense. There is no unique chemical formula and its physical properties vary widely affecting its quality and economic value. Its diversity, complexity, and behavior reflect the characteristics of the life from which it was derived and make it indeed difficult to characterize. The lack of a precise definition for petroleum makes it difficult to define oil production too. However, we believe that this difficulty has not been tackled adequately so far and that a larger effort should be undertaken, as many issues related with the oil industry could be improved if there is enough political will.

References

Carbon Materials Group (2012) Carbon history and timeline. University of Kentucky—Center for Applied Energy Research. http://www.caer.uky.edu/carbon/history/carbonhistory.shtml. Accessed 31 May 2012

Chang R (2007) Chemistry. McGraw Hill Higher Education, London

Croswell K (1996) The alchemy of the heavens: searching for meaning in the Milky Way, 1st edn. Anchor, New York

Ertl P (2002) Cheminformatics analysis of organic substituents: identification of the most common substituents, calculation of substituent properties, and automatic identification of drug-like bio-isosteric groups. J Chem Inform Comput Sci 43:374–380. doi:10.1021/ci0255782

Godwin, H. (1962). Half-life of Radiocarbon. Nature, 195(4845):984–984. doi:10.1038/195984a0

IUPAC (1993) A guide to IUPAC nomenclature of organic compounds (recommendations 1993). Blackwell Scientific Publications, Oxford

Lipkus AH, Yuan Q, Lucas KA et al (2008) Structural diversity of organic chemistry. a scaffold analysis of the CAS registry. J Org Chem 73:4443–4451. doi:10.1021/jo8001276

Meinwald J (2003) Understanding the chemistry of chemical communication: are we there yet? Proc Natl Acad Sci USA 100:14514–14516. doi:10.1073/pnas.2436168100

Bibliography

Sections 3.3 and 3.4 are based on:
Campbell CJ (2005) Oil crisis. Multi-Science Publishing Co. Ltd, Brentwood (Chapter 2)

Chapter 4
The Early Oil Industry

As we have depicted in the previous chapters, the evolution of *Homo hydrocarbonum* is not the product of an ingeniously conceived plan to master nature or some natural result of economic progress but the outcome of an intricate and remarkable process that started long before modern times. In this chapter we will take a brief look at the birth of the oil industry and the political developments that contributed first to its unfolding and later to its economic and political power. Even though oil was known by other civilizations, only the North Atlantic powers of the late nineteenth and early twentieth century attached such importance to this substance, which has maintained an unprecedented role in the geopolitical order ever since.

4.1 From Antiquity to Industrial Times

Oil and gas from surface seepages have been known since antiquity. Oil was used as mortar in early Babylon, circa 4000 BC; the ancient Chinese described the scene of natural gas seepages on the surface of lakes and swamps, 3,000 years ago; Noah's ark and Moses' basket of reeds were caulked with tar; Nebuchadnezzar's fiery furnace and the burning bush in the Bible were, it may be assumed, located on gas seepages; and the eternal flames that were worshiped by the Zoroastrians near Baku, Azerbaijan, were flammable oil-impregnated shales on the shores of the Caspian Sea, a region that keeps reappearing throughout the development of this plot and will continue to do so in the decades to come.

Almost every culture has found a use for oil. Although rarely abundant, oil was used primarily for medicinal or ritual purposes, as well as for heating and illumination. Persians, Chinese, and Romans, just to mention the most obvious examples, were exploiting oil for these purposes before the time of Christ. As late as the 1960s, Colin Campbell witnessed the rite of the Papuan natives in the highlands of New Guinea who daubed their bodies with oil collected from seepages when they gathered for their ritual sing–sing dances, a custom that had been going on since time immemorial.

C.A.S. Hall and C.A. Ramírez-Pascualli, *The First Half of the Age of Oil: An Exploration of the Work of Colin Campbell and Jean Laherrère*, SpringerBriefs in Energy, DOI 10.1007/978-1-4614-6064-0_4, © Springer Science+Business Media New York 2013

The technique for oil extraction evolved over centuries from skimming oil off the pools into which the natural seepages ran, to digging shallow pits to extract it more effectively, to early wells, to the very sophisticated techniques we use today. The early Burmese were the most advanced of their time in this regard, using bamboo pipes to case shallow wells and transport the oil. The ancient Chinese are credited with the development of a form of rig for drilling wells, consisting of a heavy stone on the end of a rope that was repeatedly raised and dropped, slowly punching a hole into the earth (Feng et al. 2012). It was the precursor of the cable-tool with which the early "modern" oil wells were drilled, nineteen centuries later. There has also been a very long history of constructing wells for water and salt brine, as salt came to be a very valuable commodity long before the Middle Ages. Even in the nine-teenth century, in the USA, wells were drilled to get salt brine; oil came as a by-product that sometimes was kept for medicine or lighting.

During this time a trade in oil was established in both Europe and the USA; there was much demand for organic compounds including derivatives from lard, whale oil, camphene, oil from seepages and coal workings which were used as domestic illuminants. In the 1850s, there was a dramatic improvement in the tech-nique for obtaining kerosene from oil—coupled with the decline in whale oil due to over-whaling—which stimulated the search for petroleum. The technique for drilling wells was already well established in North America for the production of brine, from which salt, needed for preserving meat, was extracted. There were even cases where such wells encountered gas, which was used to fuel the salt works (Yergin 2003).

The kerosene lamp by itself had been a great revolution in the way people lived, adding a usable evening to the working day, especially in rural areas. In fact, during its first 40 years, the business of the oil industry was illumination, not gasoline but kerosene. A second and greater revolution started on July 3 1886, when Carl Benz in Germany powered the first automobile with a single cylinder engine, based on the four-stroke cycle invented by Otto some years before (Eckermann 2001). At first, he used carbureted benzene distilled from coal, but soon he turned to gasoline refined from crude oil. Oil evolved from a humble by-product of salt brine to an indispens-able resource for the Western society, first for domestic illumination and finally for the fuel engines that ran the industries.

4.2   The Birth of the Oil Industry

The birth of the oil industry is generally attributed to the famous well drilled specifically for oil in 1859 by the self-styled "Colonel" Edwin L. Drake at Titusville, Pennsylvania. Nevertheless, Canadians also claim the historical parenthood of the modern oil industry: James Miller Williams initiated a production well in Ontario, in 1858, followed 3 years later by John Henry Fairbank, also in Ontario. In Europe, production statistics in Romania go back to 1854; yet others claim that F.N. Semyenov actually was the very first pioneer to dig a commercial well on the

Apsheron Peninsula, near Baku in Azerbaijan, in 1848. In fact, the Russian tradition in oil goes back one century before; in 1757, the brilliant academician Mikhail V. Lomonosov had published essentially a correct assessment of the origin of oil: "rock oil originates as tiny bodies of animals buried in the sediments which, under the influence of increased temperature and pressure acting during an unimaginably long period of time, transform into rock oil" (Campbell 2005a, p.81).

However, oil sands were mined in 1745 in Merkwiller-Pechelbronn, north-eastern France, a region with a rich mining tradition. The operation in the Pechelbronn field resembled a coal mine more than an oilfield: workers descended to the bottom of the main shaft where they dipped up the oil that dripped from the nearby galleries; they poured the oil into buckets which were lifted to the surface and taken to the refinery. It was here, in Pechelbronn, where the first drilling research was made, using a manual drill in 1813. Later, the first school of oil technology was created in the region, becoming the forerunner of the current *Institut Français du Pétrole*. In addition, Conrad Schlumberger performed the first electric logging also in Pechelbronn, in 1927.

As you can see, it is almost impossible to determine which was the very first modern oil operation ever, and in the end, it does not really matter who wears this crown. In any case the oil industry grew rapidly in the succeeding years first in Pennsylvania and along the shores of the Caspian from which it spread to many other places around the world.

4.3   From the Appalachian Boom to the Large US Companies

Drake's well, which encountered oil at a depth of 67 ft in an Upper Devonian sandstone, led to the first great exploration boom as the shallow oil reservoirs were tapped by an army of pioneers and speculators who descended on the petroleum lands of the Appalachian Basin in the USA. Stills were erected nearby to make kerosene, which, within the remarkably short span of 2 years, was being exported to Europe. Fortunes were made and lost and prices fluctuated wildly. From its outset, the industry has been plagued by "boom or bust." The reason is the special depletion pattern of oil, which flows rapidly under its own pressure from the wellbore as soon as it is tapped in a manner very different from, for example, the labor-intense process of mining coal. As a result, the market could be literally flooded in a short time by the oil coming from a few wells and then just as quickly dry up as the oil from a region was depleted.

Early explorers in Pennsylvania drilled by guesswork, although soon began to develop an empirical understanding of geology, identifying the characteristics of sites where drilling succeeded. The state of Pennsylvania appointed a geologist to investigate the nature of petroleum; his report of 1865 observed that oil tended to accumulate in anticlines, where the strata are folded into an arch, and also characterized petroleum as a "bituminous liquid resulting from the decomposition of marine and land plants."

The oil boom of the Appalachian Basin was already over by 1900. By 1885, the State Geologist of Pennsylvania stated that "the amazing exhibition of oil is only a temporary and vanishing phenomenon—one which young men will live to see come to its natural end." He was both right and wrong in his prognosis: he was right about the area he knew but wrong insofar as he did not know how much oil would be found subsequently in new areas. Now, more than a century later, we have a much better idea of that issue and can confidently repeat his words on a global scale. During the 40 years of the Appalachian oil boom, 183 oilfields had been found, yielding an ultimate recovery of 1.33 billion barrels (Gb). The Appalachian Basin, despite its early importance, is in fact quite a small province, as its total contribution would be enough to supply the world's present demand for less than a month. Nevertheless, it still has reported reserves of 28 million barrels (about 2% of the ultimate recovery), showing how reserves and production decline exponentially; old fields continue to produce a few barrels a day for a very long time during the tail end of their depletion and ever smaller fields continue to be found even in mature areas.

4.3.1 Standard Oil

In the same year that Drake drilled his well, a man by the name of J.D. Rockefeller went into partnership with a newly arrived British immigrant, Maurice Clark, to establish a trading company in Cleveland, Ohio. The Civil War of 1861–1865 created a demand for goods of all sorts so it was an opportune moment to open such a company in a place connected by two railways and the Great Lakes navigation system. Due to the boom in Pennsylvania, the new firm started to trade in kerosene and later it became profitable to enter the refinery business. They were very successful because they learned how to produce and market a uniform or "standard" product. This is how the great Standard Oil Company started, and, running on relentless business lines, it grew to become the world's largest corporation. It was the precursor of the modern company with its bureaucracy and driven solely by the motive of financial return on investment. It was basically a marketing company, seeking to control its market by both fair and foul means. It succeeded in doing so by placing a stranglehold on oil transport both by securing the pipelines and negotiating special rebates from the railways. The wild fluctuations in oil price were anathema to its orderly plans, and John Rockefeller believed that monopolies were far more efficient than competition. Even in these early years of the industry, a need for regulation had already arisen: Standard Oil was in fact exercising a function no different from that subsequently applied by the Achnacarry Agreement, the Texas Railroad Commission, or OPEC.

Standard was reluctant to enter the hurly-burly of exploration, although it eventually did so in the 1880s when another new oil province was found in Indiana. Its motive for doing so was to protect its existing market from competition, since the Pennsylvania fields were beginning to decline already. Standard Oil's ruthless capitalism was much reviled by the independent oil producers who regarded it as a creeping octopus that would eventually ensnare and devour them. It was particularly disliked in Texas and the southern states, still smarting from the Yankee victory in the Civil War. Pressure against it grew, until in 1911 the government was forced to break it up under antitrust legislation, a democratic response to an overweening monopoly that went unrivaled, except perhaps for the unswerving centralism of the Soviet Union. After the breakup, some of Standard's daughters, Esso, Chevron, Mobil, Amoco, Conoco, Sohio, and Arco—to name only the largest of the 37—grew to become some of the world's most important oil companies in their own right; three of the famous "Seven Sisters" (the seven largest oil producers of the twentieth century) were indeed Standard Oil's "daughters."

Although these companies did practice successful exploration throughout the world, in some cases pioneering new projects in the best traditions of the explorer, it may be fair to say that they owed most of their growth to their great financial strength, inherited from Rockefeller's empire which allowed them to swallow their weaker competitors. Probably the major companies have always been traders at heart, with making money being their sole objective: nothing wrong with that of course—save when one comes to consider the depletion of a finite resource that should perhaps be governed by more sophisticated criteria to better respect the value of this irreplaceable resource to humankind.

Standard Oil's control was centered on the eastern states, but new oil lands were being explored in the south. California came in first with several important discoveries before the turn of the century. The complex geology of California, where the oil occurred in strongly folded and faulted tertiary rocks, partly affected by the famous San Andreas Fault, prompted a greater attention to scientific geology here than elsewhere, and the companies in California took on increasing numbers of professional geologists to guide their efforts. Standard Oil moved in, and its affiliate, Standard of California, now Chevron, has had an exceptional reputation in exploration, in due course bringing in the giant fields of Saudi Arabia. However, Unocal was the dominant producer in California at that time. Unocal, established in 1890, managed to preserve its independence for more than one century, but in 2005, it finally merged with Chevron.

4.3.2 Texas: Spindletop and Texaco

Oil was first found in Texas in 1893, when a water well at Corsicana unexpectedly encountered oil. It was followed in 1901 by the spectacular blowout at Spindletop near Beaumont, in which 75,000 barrels a day gushed high into the sky. It was a mighty roar that heralded another oil boom, opening up a new province and enormously accelerating the industrial revolution. It transformed the economic life of the USA and was a critical contributor to its global political power. Discovery followed discovery as the new "trends" were drilled, but eventually Texas production peaked in the 1930s. Today, only a few million barrels are found each year and they are extracted from very small fields. The total endowment of Texas, resulting mainly from the early discoveries, is about 60 billion barrels (Gb), about the same as the North Sea.

The discovery at Spindletop spawned two of the Seven Sisters family, Texaco and Gulf Oil, both taken over by Chevron in 1985 and 2001, respectively. The merger of Gulf Oil with Chevron occurred in a cumbersome way, after an unsuccessful assault by a clear thinking corporate raider, T. Boone Pickens. He correctly realized that Gulf's past was worth much more to its shareholders than its future. It is a reality that most other major oil companies now face, with varying degrees of frankness. On the other side, Texaco and Chevron have a long history together that goes back to the creation of Caltex in 1936, a joint venture to develop Chevron's recent discoveries in Saudi Arabia which marked much of Texaco's evolution as an oil company.

Many of the early oil pioneers were aggressive and colorful men. The early Spindletop discovery well was drilled by a one-armed self-educated mechanic, named Patillo Higgins, who was backed by Captain A.F. Lucas, an immigrant from Yugoslavia. They eventually sold out to what became Gulf Oil. Another man on the scene was Joe Cullinan, known as "Buckskin Joe" for his abrasive manner. He set up the Texas Fuel Company in Beaumont to trade in oil and oilfield equipment, backed by New York and Chicago investors. In 1906, the name Texaco with its logo of a green "T" overlying a red Texas Star was registered. Buckskin Joe ran the business with an autocratic style that did not endear him to his investors who eventually ran him off. Some say that the company has retained a characteristically autocratic manner ever since, which perhaps owes something to its founder. Later, Texaco would be directed by Torkild Rieber, a tough Norwegian seaman who admired and supported the fascist movements in Europe, a position held by a number of other businessmen all around the world. In Spain, Rieber supplied oil and credit to General Franco in contravention of the US Neutrality Law. Later, he swapped oil for tankers built by Nazi Germany. After the beginning of the Second World War, the British intelligence uncovered German spies operating in Texaco's New York office. The captain was forced from office in 1940.

Texaco had a rather indifferent performance as an international explorer, perhaps because the oil supply delivered by Caltex, a joint venture with Chevron, would meet Texaco's needs for a long time during the past century. Like most American companies, Texaco was operating in a cutthroat environment. Growth was achieved primarily by buying up competitors; litigation was the order of the day and little time remained for exploration endeavors. On the other hand, much of the US market was controlled by Standard, so Texaco paid special attention to overseas markets. In 1936, Texaco's global marketing strategy prompted it to join forces with Chevron in the Eastern Hemisphere, creating the Caltex group. As a result, it gained access to Chevron's rights in super-rich Saudi Arabia and later to the Minas Field in Sumatra, the largest in Indonesia, which had to wait until the end of the Second World War to be developed. Thus Texaco had no need to find more oil, save for strategic reasons to reduce its dependence on Saudi Arabia; its policy for obtaining reserves seemed to be the acquisition of other companies including Seaboard in Venezuela, Trinidad Leaseholds, and Getty (the latter landing it in an expensive lawsuit with Pennzoil). After a late attempt to focus on its key domestic fields, Texaco merged into the Chevron Corporation in 1985.

4.4 Beyond America: The Caspian and the Middle East

The early development of the oil industry in the USA had a lasting world influence. The large American companies expanded overseas taking their business and technical culture with them. In technical terms, the industry's American roots have left its legacy, including, for example, its units of measurement: the traditional well-casing sizes of 95″/8 (ninety-five eighths of an inch) and 133″/8 are still in almost universal use as are such colorful drilling terms as roughneck, rat-hole, and kelly bushing, not to mention a piece of equipment delightfully known as a "donkey's dick." But the USA was not by any means the only pioneering oil country. During the nineteenth century, there were already developments in many places including Baku in Russia, Ontario in Canada, Borneo, Burma, Sumatra, Romania, Poland, Trinidad, Peru, and Mexico.

4.4.1 The Shores of the Caspian

Of all the oil-rich provinces, the most important was Baku, a backward and poorly administered territory on the shores of the Caspian, on the southern fringe of the Russian Empire, in today's Azerbaijan. Oil extracted from hand-dug pits had been a state monopoly, but in the early 1870s, the area was opened up to private capital. One of the first entrepreneurs to arrive on the scene was Ludwig Nobel, a member of the inventive Swedish family, which made a fortune out of dynamite. The family is now remembered by the Nobel Prize, which it endowed. Since there was no sufficient oak to make traditional wooden barrels, the Nobels developed the world's first tanker, the Zoroaster, to transport oil through the Caspian.

The Rothschild bank came in later to finance a railway to Batumi on the Black Sea, opening an export route to the West. They in turn were followed by a new company by the name of Shell, which started exporting Baku kerosene in tankers to Europe and the Far East. Hastened by the growing power of Standard Oil in the USA, Shell merged with Royal Dutch in 1907, which had been pioneering oil production in the Dutch East Indies (now Indonesia) to become the giant Royal Dutch/Shell or simply Shell, as it is known today.

The Baku oilfields lie in a Tertiary basin in front of the Caucasus. The geology is characterized by complex folds and faults and multiple reservoirs. Seepages of both oil and gas were abundant. A peculiarity of the geology was the presence of numerous so-called mud volcanoes, as also found in Romania, Colombia, and Trinidad. These volcanoes are mounds of mud, up to several hundred feet in height, which form over active gas seepages. They explode and catch fire sometimes. Without doubt, most of the fields were found by hit-or-miss. It was evidently easier to hit than to miss in this prolific area that in 1900 produced as much as 75 million barrels from 1,700 wells less than 1,000 ft deep. It was a violent place of banditry and strife, with appalling operating conditions. No less a figure than Joseph Stalin was a workers' leader, masterminding strikes and disturbances in Baku in the early years of the century. Such pressures spread and culminated in the Bolshevik rising of 1917, which, to put it mildly, transformed the world's political scenery. It was not to be the last occasion on which oil shaped human destiny, the most important of which is about to come. The Bolshevik Revolution effectively brought to a close the first Caspian oil boom. Hitler tried to capture Baku during World War II. He knew it was Azerbaijani oil what fueled the Red Army, and the Germans were running out of oil quickly (Yergin 2003). Had the Nazis succeeded in that mission, the outcome of the war may have been different.

Inheritors of the Russian tradition, the Soviets became very efficient explorers. Besides, they were able to approach their task in a scientific manner, being able to drill holes to gather critical information, whereas their Western counterparts had to pretend that every borehole had a good chance of finding oil. In the years following World War II, the Soviets found and brought into production the major oil provinces of the USSR, finding most of the giant fields within them; Baku had become a mature province of secondary importance. The Soviet Union had ample onshore supplies, which meant that it had no particular incentive to invest in offshore drilling equipment. The Caspian itself was largely left fallow, although the borderlands were thoroughly investigated. Of particular importance was the discovery of the Tengiz field in 1979 in the prolific pre-Caspian basin of Kazakhstan, only 70 km from the shore. The problem was that the oil had a sulfur content of as much as 16%, calling for high-quality steel pipe and equipment, not then available to the Soviets. Development was accordingly postponed. The fall of the Soviet regime in 1991 opened the region to Western investment.

BP took a pioneering role with Statoil, its Norwegian partner. Interest was at first aimed at the offshore extensions of the Baku trend, where a number of prospects, already identified by the Soviets, were successfully tested, finding some three billion barrels (Gb) of oil, which, however, was not enough to have any particular world significance; meanwhile, ExxonMobil had withdrawn from Azerbaijan altogether. It is unlikely that more than three billion barrels (Gb) remain to be found in the country.

Soon Kazakhstan also attracted interest. Chevron–Texaco, together with ExxonMobil, agreed to develop the Tengiz field. One of the problems has been the disposal of the huge amounts of sulfur that had to be removed from the oil by processing. Plans to increase production in the field are now shelved.

The greatest interest of all, however, was attached to a giant project named Kashagan, which was identified in the shallow waters of the northern Caspian off Kazakhstan, leading to the entry of a European consortium comprised of BP-Statoil, Agip, British Gas, and Total. The enthusiasm waned as the companies began to get into the details. The reservoir was composed of individual separated reefs and the integrity of the salt seal seemed weak in some parts of the structure. Operational challenges were also monumental: the waters were shallow, making it difficult to bring in and position equipment, while also posing environmental threats to the breeding grounds for sturgeon shoals supporting the Russian caviar fisheries. If that was not enough, a gruesome, chilling wind blows in winter, covering everything in ice. Nevertheless, the companies succeeded in drilling three testing boreholes at an astronomical cost, announcing that they had found between 9 and 13 billion barrels

(Gb). BP-Statoil decided to withdraw. In addition to these main projects, the Russian themselves have made a two billion barrel (Gb) discovery in the northwest part of the Caspian, and Turkmenistan has announced an oil discovery of uncertain size off its mainly gas-prone territory.

In short, it is now clear that the Caspian has been a great disappointment. Total reserves for the offshore probably stand at about 25 billion barrels (Gb), with new exploration offering potential to perhaps another five, a good deal less than the 44 billion barrels (Gb) mean estimate proposed in a study by the USGS in 2000, and much less than the 200 billion barrels (Gb) announced in some Western newspapers. Even in the unlikely event that the USA had exclusive call upon it, the Caspian off-shore could provide only 10% of its needs for only a relatively few years.

4.4.2 Persia: The First Sample of the Middle East

The greatest oil province of all, the Middle East, was also attracting attention as the nineteenth century drew to a close, but to do business there was difficult. Most of the area was controlled by the Ottoman Empire as late as the first decades of the twentieth century, with its decadent and corrupt Sultan in a harem surrounded by eunuchs. The rest was in the hands of the Shah of Persia, today Iran, whose authority barely extended beyond his own capital. The Germans became interested in building a railway from Berlin to Baghdad as part of a foreign policy initiative aimed to catch up with the colonial expansion of France and Britain in other parts of the world. As a land power, it recognized the military mobility afforded by railways, which it conceived would be more effective than the slower British sea power. It secured to this end a concession in Anatolia and Mesopotamia, which included mineral rights for 20 km on either side of the track, presumably as a source of building stone. Its engineers soon reported the numerous oil seepages that they encountered in the Mosul area of what is now Iraq. Although the Sultan was alerted about the possibilities and tried to retract the rights from the German railway company, he was too idle and ill informed to do much about it.

At about the same time, the head of Persian customs, General Antoine Kitabgi, hearing of the growing oil interest in the vicinity, resolved to see if he could let an oil concession into his country. After one or two false starts, he managed in 1900 to bring it to the attention of William Knox D'Arcy, an entrepreneur who had just returned to London from Australia where he had made a fortune in gold mining. He saw the possibilities and eventually secured the rights: a £20,000 signature bonus being the main inducement to the impoverished Shah.

However, the oil was located in the territory of the Bakhtiari tribe, and drilling permission had to be obtained from the Bakhtiari Khans in addition to the concession of the Shah; throughout their relation with the British, the Khans proved to be intelligent negotiators. After reaching an agreement with them, drilling commenced under appalling conditions: donkeys and mules were used to carry heavy equipment, no roads or no port facilities existed, pipes had to be laid and maintained over mountainous territories with summer temperatures of over 40°C (110°F), and success was slow in coming. D'Arcy was becoming overextended but was encouraged when the British government started to take an interest in his project. Britain had always sought to deter Russian expansion into the Middle East in order to protect its trade route to India and the Far East, and oil was now a new factor. D'Arcy's immediate problems were, however, partly solved only when the old established Burmah Oil Company, based in Glasgow, agreed to take a share in his company.

In January 1908, after six long years of travail and disappointment, the third well was spudded at Masjid-i-Sulaiman (the Mosque of Solomon) in the Zagros foothills: it was pretty much the last throw of the dice. By May, the well was down to 1,000 ft without results, and a cable to suspend operations was received from London. The local manager, G.B. Reynolds, however, decided to continue until he received written confirmation. His initiative was rewarded around 4 a.m. on the morning of May 26th, when the well blew out throwing a jet of oil fifty feet into the air.

In world resource terms, it was a climactic event. The world contains no more than about 30 significant petroleum systems with the unique set of geological circumstances to yield prolific oil, and this discovery in 1908 had found the largest. It was undoubtedly a turning point in history that gave birth to the Bakhtiari Oil Company in 1909, which immediately became the Anglo-Persian Oil Company (APOC), later Anglo-Iranian Oil Company (AIOC) in 1935, and finally, British Petroleum (BP) in 1954.

4.5   World War I: From Horses to Air Fighters

Another turning point of a different sort was about to unfold in 1914: the First World War. Britain's last major naval engagement had been the Battle of Trafalgar in 1805, a critical action in the Napoleonic wars. At the height of Empire, the British Navy was the corner stone of Britain's power, but by the turn of the century it had become more of a symbol, with polished brass, holystoned decks, and smartly dressed crews, than an efficient fighting machine. It was just this pageantry what may have impressed the mercurial character of Kaiser Wilhelm II, himself a grandson of Queen Victoria and honorary admiral in the British Navy, when he came to take part in his favorite sport of yacht racing at Cowes. Thus, by the late 1890s, Germany had launched a full-scale challenge for naval supremacy against England. And so began the Anglo-German arms race, as each new German warship had to be matched by a British one to maintain the balance of power. Gradually the emphasis changed from the pomp and splendor of the marine band on the quarterdeck to actually making the thing a lethal weapon, able to outspeed and outgun its competitor. The maverick Admiral Fisher was dedicated to this transformation. He realized that his ships would have to convert to oil fuel to obtain the performance he expected, but his proposals met resistance. For the first time, but certainly not the last, the issue was security of oil supply: Britain had no oil of its own and was reluctant to rely on

American oil or even on Shell oil with its Dutch connection. What England needed was its very own controlled supply of oil. Winston Churchill, then the First Lord of the Admiralty, concluded that Persia was the answer for two reasons: first, Persia had oil to supply the Navy; and second, a British presence in the Middle East would deter the threat of German or Russian expansion in that area. Thus, the government took up a 51% interest in the Anglo-Persian Oil Company; the royal assent was granted 6 days before war broke out.

The war opened with cavalry charges, as plumed Uhlan lancers galloped into action, and steam engines to mobilize troops along the rails, but it ended with tanks and airplanes driven by internal combustion engines that ran on fuel refined from crude oil. Oil became the great new driving force of the world, changing the meaning of the term "horsepower." But in terms of oil resources, perhaps the most significant feature of the First World War was that Turkey backed the losing side. Had that country been an ally or neutral, events would have turned out very differently. Turkey controlled the area that we know today as the Middle East, inhabited by multiple, dispersed Arab tribes. Therefore, Britain had a motive to encourage Arab nationalism, which effectively resulted in the breakup of the Ottoman Empire into new administrations after the war, of which Iraq, Kuwait, and Saudi Arabia are among the most important in terms of hydrocarbon reserves. This breakdown was to have far-reaching economic and political consequences that have still to be played out. Not far below the surface was the division of the region's oil rights to the three victorious allies, Britain, France, and the USA. Thus, the West came to exert wide political control over the region, that otherwise may have remained under the sphere of the Asian powers.

4.6   The Development of the Middle East

The first solution for the division of oil rights among the victorious powers was to share them. This was achieved by the formation during the 1920s of the Iraq Petroleum Company, owned by Shell, BP, Compagnie Francais des Pétroles (CFP, now Total), Mobil, and Esso, not to forget the legendary Calouste Gulbenkian, who got 5% as payment for putting the deal together. This group had what is called an Area of Mutual Interest (AMI) agreement. This agreement prohibited any independent activity undertaken by any and each of the partners in the area of the former Ottoman Empire. It became known as the Red-Line Agreement, covering all the productive Middle East territories outside Iran and Kuwait, and was the cause of bitter conflict for a long time.

Chevron, which was not restricted by the Red-Line Agreement, took up rights in Bahrain in 1929 and struck oil 2 years later. This find, coming from Tertiary sandstones at fairly shallow depth, was itself comparatively modest, but it was nevertheless of immense importance, for Bahrain lay only a few miles off the coast of Saudi Arabia. Up to that point, interest in oil had been concentrated on the huge folded structures of the Zagros Foothills in Iran and Iraq that were obvious surface features visible for miles around. Many geologists, seeking analogues for this familiar type of prospect, were then skeptical of the "platform province" to the west of the Persian Gulf, where the strata were largely obscured below sand dunes and, where seen, were flat-lying or, at most, shallow dipping. At first sight, it seemed to lack adequate structure to provide large traps for oil. So, the discovery of oil in Bahrain on the edge of this new province carried immense implications, which were at once recognized. Chevron began negotiating for rights in Saudi Arabia, partly through an eccentric and disaffected Englishman by the name of Harry St. John Philby. He was trading in Jidda and was, remarkably enough, no less than the father of the infamous British double agent, Kim Philby.

King Ibn Saud, himself a British protégé from the War, was desperately short of money, and Chevron clinched the deal in 1933 with delivery of 35,000 gold sovereigns that were shipped to Arabia in seven boxes aboard a liner that belonged to the famous Peninsular and Oriental Steam Navigation Company (P&O) based in London. It was a substantial and risky investment at the time, for no one could have imagined that Saudi Arabia would become the world's most prodigious oil province, with an ultimate endowment of about 300 billion barrels (Gb), 16% of the world's total. Chevron, when it later found that it lacked the resources to develop the area single handedly, brought in Texaco, followed in 1947 by Mobil and Esso, the latter two in flagrant disregard of the famous Red-Line Agreement. This grouping formed the Arabian–American Oil Company (ARAMCO), the emphasis being on the second word. In prewar days, Britain under its imperial mantle had successfully exerted an almost exclusive influence throughout the Middle East, but in its weakened and socialist postwar state had loosened its grip in favor of the USA. Ibn Saud, absolute ruler of a feudal and primitive country that was little more than his private estate, effectively became an American satrap. The further evolution of this remarkable and extraordinary situation has yet to unfold with or without the House of Saud. We will return to the issue in later chapters.

While rights to Saudi Arabia were being negotiated, BP and Gulf turned attention to Kuwait, which lay also on the western shore of the Persian Gulf and outside the Red-Line Agreement. They eventually decided to join forces rather than compete for the territory and signed a lease for it in 1933. This agreement completed the primary carve-up of the Middle Eastern oil provinces by European and US entities and began their fateful serious financial interests in the region.

Although by far the most important, the Middle East was by no means the only oil territory being explored and developed. Exploration had expanded throughout the world, such that most of the world's onshore oil basins, and many of the giant fields within them had been identified prior to the Second World War. Most progress was in the Western Hemisphere, especially in the USA itself, which was already becoming a fairly mature province, but also in Venezuela and Mexico, where impressive finds were made. Shell, which had rather missed out on the carve-up of the Middle East, took up a strong position in the Western Hemisphere, competing successfully with the major American companies. Generally smaller scale operations were also taking place in many other countries. By 1935, 25 were in production. The USA was producing 64% of the world's oil needs. The Middle East was barely represented: Iran, the largest producer, was in seventh place with only 2%. The other countries were, in decreasing order: India (including Pakistan), Poland, Peru, Colombia, Argentina, Trinidad, Japan, Sarawak, Brunei, Iraq, Canada, Germany, Egypt, Sakhalin, Ecuador, France, Italy, Czechoslovakia, and Bolivia. What a different world it was!

Seven major companies, comprising Shell, BP, Esso, Mobil, Chevron, Texaco, and Gulf, later dubbed the "Seven Sisters" by Enrico Mattei, the Italian oilman who built ENI, had already brought world supply under their control. With most oil coming from the USA, security of the supply was not a serious issue, although one not altogether without concern. The Soviet Union was closed to foreign companies, and those with rights from the prewar days in Baku were formally expropriated in 1928. Mexico also ousted the foreign companies 10 years later as a nationalist movement, somewhat akin to the populist movements of South America, gained political ascendancy, believing that foreign influences were becoming excessive, perhaps with some justification. Its oil industry was placed in the hands of a state enterprise PEMEX, the first example of the state oil company that was later to be copied widely. These expropriations were harbingers of what was to come, for by 2010 most of the world's oil production and reserves reside within national oil companies.

References

Eckermann E (2001) World history of the automobile. SAE International, Warrendale
Feng, L., Hu, Y., Hall, C. A. S., & Wang, J. (2012). The Chinese Oil Industry: History and Future (2013th ed.). Springer.
Yergin D (2003) The prize: the epic quest for oil, money, and power, 1st trade paperback edn. Free Press, New York

Bibliography

This chapter is primarily based on the following sources:
Campbell CJ (2005a) Oil crisis. Multi-Science Publishing Co. Ltd., Brentwood

The technical and economic prospects in Sect. 4.4.1 are based on
Campbell CJ (2005b) The Caspian chimera. In: Newman S, McKillop A (eds) The final energy crisis. Pluto, Ann Arbor/London

Chapter 5
King Hubbert: A Pioneer of a Different Kind

*Being outspokenly correct when the conventional wisdom would
have it otherwise may not win popularity contests, but the
vitality and intellectual integrity of men such as King Hubbert
are rare and precious qualities. Recognition of King Hubbert
marks our great gratitude and humble respect for all that he
has done for our science and for this country.*

–Barry Raleigh, Director of the Lamont-Doherty Earth Observatory
during the 1981 Vetlesen Prize ceremony, the highest award in
Earth Sciences

*The importance of any science, socially, is its effect on what
people think and what they do.*

–M. King Hubbert

In 1950, world oil production stood at 10 million barrels a day (Mb/d), but within
20 years, it had risen to 45 Mb/d, a staggering near fivefold increase. It was in this
environment that M. King Hubbert delivered a lecture that challenged the optimism
that prevailed at the time in the oil industry. Although it may seem to be a golden epoch
for the oil companies, there were darkening shadows about the future of the industry.
The rapid growth was in a sense a reaction to a new uncertainty and insecurity. The
companies had no intrinsic reason to flood the world with cheap oil, which, had they
been assured of their future, would have been contrary to their long-term interests. The
nationalization of the oil industry in the Soviet Union in 1928 and in Mexico 10 years
later was only a ripple compared to the storm that was in formation.

5.1 The Oil Environment from the 1950s to the 1970s

Even before the Mexican nationalization in 1938, large companies had relocated
their main operations in Venezuela, where new oil fields were discovered. The
Venezuelans pushed hard for higher taxes and royalties, and the companies, not

C.A.S. Hall and C.A. Ramírez-Pascualli, *The First Half of the Age of Oil: An Exploration
of the Work of Colin Campbell and Jean Laherrère*, SpringerBriefs in Energy,
DOI 10.1007/978-1-4614-6064-0_5, © Springer Science+Business Media New York 2013

wanting to risk their investments in the country as they did in Mexico, cooperated with the government this time. Thus, a new agreement was reached under the principle of a "fifty–fifty" share. Aramco and the Saudis signed a similar agreement in 1950. At the time, Iran had been in dispute with BP over the level of government take. The Iranians found inequitable the share: they should receive royalties of £90 million, while the company registered a profit of £250 million. Half of the profits went to the British government that held 51% of the shares. Complaints about foreign exploitation rang in the Iranian Parliament and in the streets as well. The most outspoken and impassioned voice was that of Mohammed Mossadegh, a controversial, frail-looking, 70-year-old, land-owning aristocrat, who harangued Parliament with theatrical speeches. The situation deteriorated when the prime minister, who proposed moderation, was assassinated, followed soon by the Minister of Education, who suffered the same fate. Events were spiraling out of control. In the face of these popular pressures, the Parliament voted for the nationalization of BP's Iranian affiliate. On April 28th 1951, Mossadegh was elected as prime minister to implement the decision. By September, a British warship, with the band playing "Colonel Bogey," had evacuated the last British nationals from Abadan, the company's base since 1908.

It was a far cry from the gunboat diplomacy of an earlier epoch. A war-weakened England no longer had the stomach for empire or the enforcement of contract, although the British did set up commercial embargoes and diplomatic pressures against Iranian oil and anyone involved in its exploitation and commercialization. Besides, the socialist government in Britain was then bent on nationalizing almost everything in sight. So, who should deny the Iranians the right to do the same thing? The British retreat from Iran was much welcomed by the Americans, perhaps for anticolonialist reasons, but as likely with an eye to oil. A few years later, when the Iranian crisis was resolved, American companies found themselves occupying, through their stake in the consortium that had replaced BP, much of the position once held exclusively by British Petroleum.

BP's position changed from a secure and highly profitable business to a risk-taking investor. After the Iranian crisis, BP had to start searching other sources, launching a vigorous and remarkably successful exploration campaign that literally reshaped the world. Their quest for new supplies culminated with two major oil fields: Alaska, discovered by the Atlantic Richfield Company (ARCO) but owned majorly by BP, and the North Sea. At the time, the Iranian crisis was seen as little more than an unfortunate chapter in a changing postwar world. But in fact, it had far-reaching implications and led to shifts of attitude and policy, some of whose consequences are yet to be played out in the Middle East and elsewhere, as the dubious performance of the company during the recent oil spill in the Gulf of Mexico questioned the viability of deepwater exploitation worldwide.

Before 1951, BP had a concession that ran for 60 years from 1901. It relied on Iran for most of the oil it needed to supply its European and imperial markets that were gradually growing. It could easily balance supply and demand, and make long-term plans, both for the duration of the concession and with the reasonable expectation that it could be extended. In short, the company knew what resources it had and could plan an orderly exploitation so as to conserve them in a reasonable way. Such an attitude likewise influenced the American companies in their home country, where they owned the mineral rights outright. It had been an epoch of stability—even complacency, one could argue—but at least it provided no motive for squandering resources.

Other major companies who relied on Middle-East oil came to realize that they were becoming increasingly unwelcome tenants of unfriendly landlords: an ironic contrast with the socialist attitude to property in Europe, where the former was favored at the expense of the latter. The companies concluded that they had only two priorities: to produce as fast as possible while they still owned the rights and to find more sources of oil to lessen their dependence on a single supply.

High production rates meant the companies had to find new markets, and the main challenge of the epoch was to do just that, dumping cheap oil on the world. It soon created an energy-dependent society, driving to work from suburbia and buying consumer durables that could be transported around the world at minimal cost by cheap oil. Strawberries became available everywhere on every day of the year. The European market in particular was opened up, even to the extent of fueling electricity generation by cheap oil imports at the expense of indigenous coal.

To find new sources of cheap oil was a greater challenge. It was obvious already that nowhere in the world could be compared with the abundant Middle East. Nevertheless, for strategic reasons, as opposed to strictly economic ones, exploration was stepped up. Attention turned to Africa, a continent that had not been widely explored before. BP and Shell in joint ventures took up pioneering positions in East and West Africa, the latter soon yielding important discoveries in Nigeria. French companies turned to Algeria, where they were rewarded by the giant Hassi Messaoud discovery in 1956 and Hassi R'Mel in 1957, the largest oil field and gas field in Africa, respectively, a venture where Jean Laherrère was directly involved. Both European and American companies developed a third prolific new basin in Libya, with major discoveries in 1958 and 1959.

Other findings were made in India, Indonesia, Australia, and Latin America. Meanwhile, behind the Iron Curtain, the systematic exploration of the Soviets was being rewarded by major discoveries of oil and gas. Finally, in 1968, towards the end of the epoch, another frontier was crossed with the discovery of the giant Prudhoe Bay Field in Alaska, which was of enormous importance to the United States, giving it a second lease of life after discovery onshore in the lower 48 states had peaked. The field was found by ARCO who had leased the crest of the structure which contained the gas cap, but BP's lesser bid yielded the flanks where most of the oil was. BP bought a lot of acreage on the so-called North Slope and then drilled several dry holes on the Colville structure. The Sinclair Company (and ARCO, which bought it in 1969) did succeed in making a unitized deal for drilling Prudhoe Bay. From April 22, 1967, to June 24, 1968, ARCO drilled the structure. It was the last chance for the North Slope, and it was a success. However, BP was the largest contributor of lands on the flanks (bought at 17.80 $/acre) and got the operatorship of the field. Prudhoe Bay did much to compensate BP for the loss of Iran. It was no mean feat for the company, a newcomer from Europe, to walk off with most of North America's largest oil field.

5.1.1 Offshore Exploration

These new areas were onshore, but attention also began to turn offshore. In fact, offshore extensions to fields had been attacked already by drilling angled holes from the land, as in Trinidad and Peru. Some shallow water prospects had been drilled from steel platforms erected on the seabed, as in Lake Maracaibo in Venezuela and the Gulf of Mexico. What was needed, however, was a mobile floating rig that could be used for truly exploration purposes where the cost of a fixed platform could not be justified. Engineering work in this direction launched into operation the first such floating rig, the Breton Rig 20, designed by John T. Hayward, in the Gulf of Mexico in 1949. It consisted of no more than a barge with a drilling rig mounted upon it: the innovations being in the seabed wellheads and the connections with the barge above. This technology could be used in waters up to about 100 m in depth but was very susceptible to wave conditions.

Further development led to the concept of the semisubmersible rig, which was based on the ingenious idea of building the rig on two submerged pontoons that floated below the wave base, providing a stable platform irrespective of surface conditions. The first such rig, Blue Water No. 1, came into operation in 1962. This new technology widened the scope of exploration to the continental shelves of the world within water depths of about 200 m (650 ft). The fleet of such rigs rapidly grew so that during the 1960s semisubmersible drilling was undertaken in Borneo, Iran, Canada, West Africa, the United Kingdom, Norway, and New Zealand. Designs were continuously improved; by the end of the epoch, wells were being routinely drilled in the stormy North Sea.

Two alternative technologies deserve mention. One was the jack-up, in which the platform, having arrived on location, put down long retractable legs that sat on the seabed, and jacked itself up above the waves. The Zapata Company of Houston dominated this market. It was managed by George H. Bush, who later would become the American president; this experience allowed him to gain a particular insight into the oil business that was later to influence American foreign policy. The second alternative was the drill ship in which a drilling rig was mounted amidships on a conventional vessel, which was held in place either by anchors or thrusters. Global Marine of California pioneered this approach to meet the deeper water needs of that State.

Much play is made of deepwater exploration, which some hold will provide a solution to the actual supply crisis. The technology of the semisubmersible rig did indeed make a major contribution by bringing the reserves of the world's continental shelves into reach for the first time, but we think it will prove to be the last technological breakthrough having a significant global impact. Most subsequent technology has succeeded mainly in increasing production rate and so accelerating depletion, without adding much in terms of reserves.

5.2 Hubbert's Prediction

In this whirling world of new discoveries and growing production, not many people — inside and outside the oil industry — were worried about an obvious fact: oil is a nonrenewable resource, therefore the party will be over at some point in the future. This statement has little intellectual merit indeed, since it is nothing but a tautology; what is worth of the highest appraisal is an accurate forecast about when it is likely that supply will not be able to continue growing, based on a thorough understanding of the dynamics of discovery, production, and depletion. It is still more meritorious if such a forecast is done in the face of a growing industry that is spreading worldwide. In 1956, Marion King Hubbert made such a prediction about oil extraction for the USA. In the remainder of this chapter, we will delve into his analysis. This part of the book is based mostly on our own research, although many ideas of Laherrère are mingled along the text.

Hubbert based his calculations on figures published by L.G. Weeks, a geologist at Standard Oil, New Jersey, who had estimated the "total amount of crude oil that could reasonably be expected to be produced by productive methods, and under economic conditions prevailing in 1947" within the land area of the USA, excluding Alaska (Hubbert 1962 explains the estimates used in detail). For the offshore, he used estimations published by the USGS and O.P. Jenkins, from the California Division of Mines (Hubbert 1956).

Hubbert was a distinguished scientist who had published a number of important articles on many aspects of petroleum geology including a remarkable interest in the social implications of resource depletion. He was working at the renowned Shell's Exploration and Production Research Division when he published two predictions about the year when US production would likely peak. It is worth to remember that Hubbert was trying to chart the path of production followed by the oil industry in order to address the issue of energy security at the national and international levels (Hubbert 1956). Hence, the title of his lecture was related to the role of nuclear energy as a substitute for all fossil fuels rather than a mere technique or a forecast for oil production in the US lower 48 states (USL48). In 1949, he had stated already that the fossil fuel era would be of very short duration in geological terms. "Nuclear Energy and the Fossil Fuels" was presented in March 1956 at a meeting of the American Petroleum Industry. The story says that 5 min before the presentation, a Shell executive was on the phone, asking him to withdraw his forecasts, but Hubbert was too stubborn to yield.

What M. King Hubbert had in mind was the development of a technique suitable for the analysis of all fossil fuels and other nonrenewable resources. His analyses were based on two basic assumptions:

1. The rate of production (measured in barrels per year in the case of oil) starts at zero, grows, and declines to zero again as the resource becomes depleted, after "passing through one or several maxima."
2. If we plot the rate of production year by year, the cumulative production is equivalent to the area beneath the resulting curve. This cumulative production must be less than or equal to the ultimate reserves of oil (or other nonrenewable resource).

The part of the curve where the production rate grows from zero was already solved by the existence of annual production records; in the case of oil for USL48, the cumulative production — the area under the production curve — was 50 billion barrels (Gb) up to 1956. On the other hand, the production rate must decline to zero over a prolonged time; for oil, this rationale can be justified by "the slowing rates of extraction from depleting reservoirs" empirically observed (recall the oil wells in Pennsylvania, Sect. 4.3). Thus, we have an idea about how the two extremes of the oil curve should look like: an exponential increase during the growing period — according to the existing records — and an exponential decrease during the decline — due to the slow rates observed in old reservoirs (Fig. 5.1).

What can be said about the middle part where the maximum production rate is attained? Since the oil production rate was still growing in 1956, Hubbert considered that the maximum rate should be greater than the rate already reached at the time, but how much? Hubbert built a straightforward argument using two estimates of oil reserves for USL48: 150 and 200 billion barrels (Gb). By 1956, 50 billion barrels (Gb) were already produced; given the form of the curve described by Hubbert and using the first estimate, it was "impossible to delay the peak for more than a few years and still allow time for the unavoidable prolonged period of decline [...] the curve must culminate at about [the year] 1965 and then must decline at a rate comparable to its earlier rate of growth" (Hubbert 1956, p. 23). Hubbert only filled the gap with a simple curve showing one peak. Assuming any other shape would have been unjustified. Without computers available, his curve was drawn by hand without specifying any equation to describe the model, and the area below the curve was estimated by counting the squares in the graph, as he explicitly stated:

> The unit rectangle in this case represents 25 billion barrels so that if the ultimate potential production is 150 billion barrels, then the graph can encompass but six rectangles before returning to zero (Hubbert 1956, p. 22).

Thus, the dynamics of a complex physical problem were abstracted to the area under the curve in a chart using basic concepts of integral calculus, not a simple achievement indeed if we consider that, as late as 1949, mathematics, physics, and chemistry were not standard courses in the geology curriculum (Clark 1983).

The second estimate for reserves, 200 billion barrels (Gb), was used as a form to verify the sensitivity of the main result. Hubbert wanted to show that 50 billion barrels (Gb) more — about eight times the oil in the East Texas oil field, the largest in the USL48 — did not delayed the peak more than 5 years. Thus, by contrasting both predictions, we obtain a fuller perspective about oil depletion: adding a significant amount of reserves delays the production peak only years or maybe decades, but certainly not centuries, and does not imply that the production rate will increase significantly. However, this second estimate happened to produce the celebrated forecast that USL48 oil production would reach its maximum in 1970. Even if Hubbert had published only his main estimation — which situated the peak in 1965 — 5 years is a very acceptable margin of uncertainty given the available data and all the uncertainties involved.

5.2.1 Technological Change and Discovery Cycles in Hubbert's Lecture in 1956

Another issue discussed by Hubbert is the improvement of extraction techniques, what economists like to call "technological change." The estimates for reserves

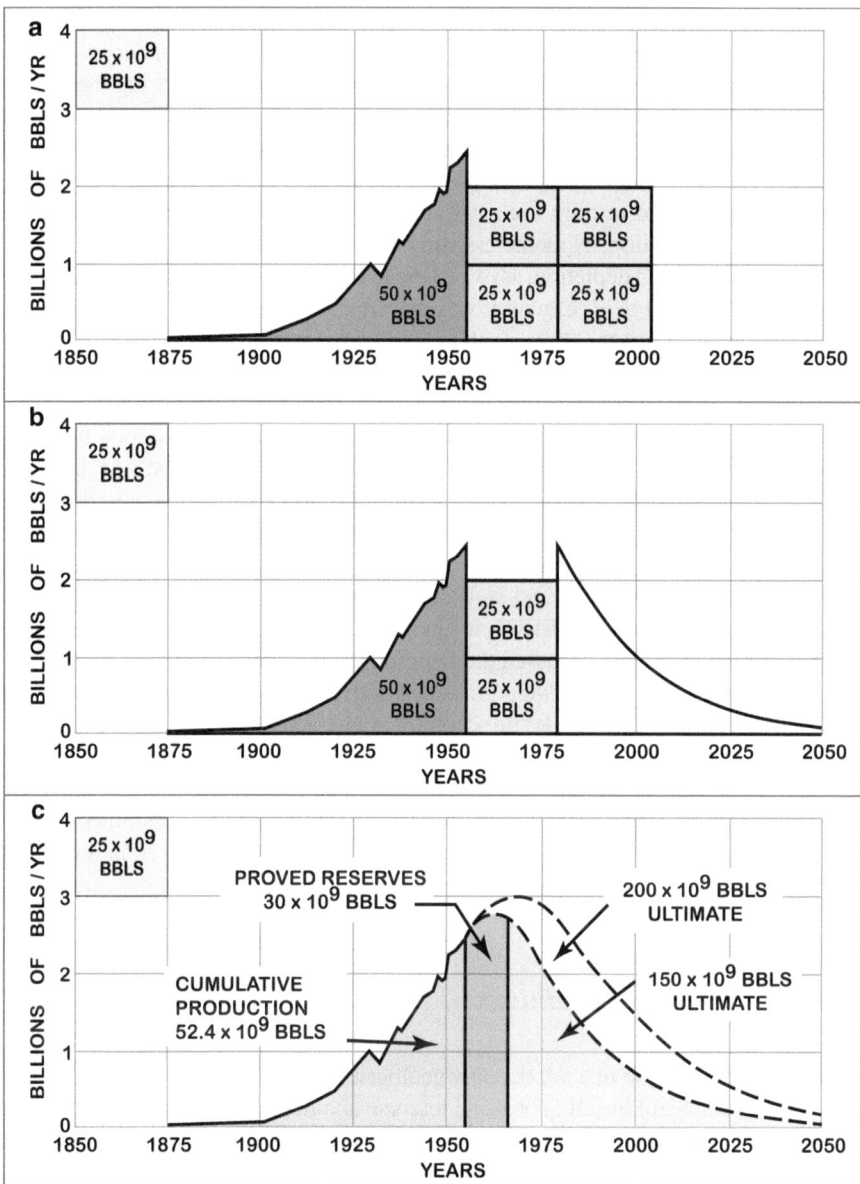

Fig. 5.1 A graphic representation of the analysis proposed by Hubbert (1956); (**a**) past oil production in the USA in 1956 had reached around 50 Gb; with an ultimate recoverable of 150 Gb, there remained 100 Gb to be produced; (**b**) the decline side could be expected to be similar to the growth side, so there remained 50 Gb in between; (**c**) with these considerations, the curve could not be very different from the figure that Hubbert published actually in his lecture in 1956

considered only the amount of oil that could be recovered by the technology available at the time. The most important "new" technology was secondary recovery. According to Hubbert, the results in the 1950s were far from increasing the rate of production and only had the effect of reducing the rate of decline after the peak; in other words, secondary recovery could neither delay the peak nor produce a second maximum but could only enlarge the base of the curve (Fig. 5.1; see Hubbert 1956). Today, secondary and tertiary recovery techniques are able to increase the maximum rate of production. Nonetheless, this increase is achieved, arguably, at the expense of accelerated depletion, as experienced in the well-known case of Cantarell, the Mexican giant oil field. Cantarell was placed under a technique involving nitrogen injection in the year 2000, reaching the highest production rate in Mexican history 3 years later. At the time, only Saudi Arabia's Ghawar, the largest oil field worldwide ever, surpassed Cantarell's production rate. This historical rate was followed by an also historical decline that left the Mexicans with serious financial issues at the national level whose complete consequences remain to be seen (Fig. 5.2). Additionally, secondary and tertiary recovery techniques require huge amounts of energy to operate, so the net yield of energy available to society—the energy obtained from the reservoir less the energy invested in the recovering technique—is lower Murphy and Hall 2011. We will return to the significance of energy return on investment (EROI) in Chap. 9.

M. King Hubbert also discussed the importance of discovery periods by examining the case of Illinois where two periods of discovery occurred; the first period was based on surface geology, while the second was characterized by the use of seismic techniques (Hubbert 1956, pp. 11–12). This pattern yielded a production profile with two peaks. However, the USL48 were thoroughly explored by 1956, and only small discoveries could be expected to be realized. Thus, the possibility of any future discoveries significant enough to yield a higher production peak in the future was unjustified.

5.2.2 Other Expert Opinions

Hubbert was neither the first nor the only geologist to forecast such a peak around this date. In fact, Hubbert himself quotes and refers to another study, the Chase Manhattan report titled "Future Growth and Financial Requirements of the World Petroleum Industry" by Joseph E. Pogue and Kenneth E. Hill that was published on February 21, 1956, a few weeks earlier than Hubbert's work. This report was presented at the annual meeting of the Petroleum Branch of the American Institute of Mining, Metallurgical, and Petroleum Engineers and was covered in the New York Times. The report concluded that the peak of production for the USL48 would occur likely between 1965 and 1970. Pogue and Hill based their forecast on the assumption that only 85 billion barrels (Gb) of oil would be discovered in the lower 48 states after 1956, yielding an ultimate of 165 billion barrels (Gb). Jean Laherrère's estimate for the ultimate yield of the US falls between 200 and 230 billion barrels (Gb) (Laherrère, personal communication).

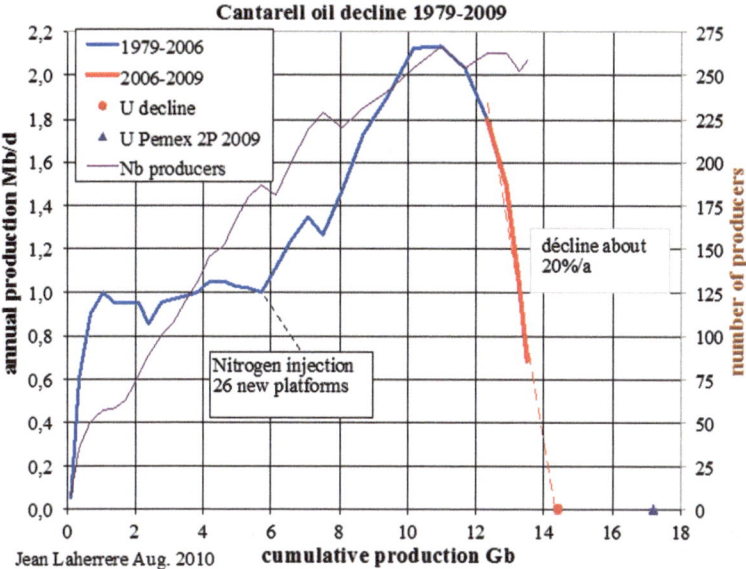

Fig. 5.2 Oil production and number or producing wells (Nb producers) of the Cantarell complex of fields in Mexico with estimations of ultimate recoverable from Jean Laherrère and Pemex. After having reached the second highest production worldwide in December 2003, production declined at rates near 20%; Mexico used oil revenues to repay its debt in 2003

With all these issues in mind, it is evident that the global picture is far more complicated than the image derived from the lower 48. In the same study, King Hubbert used 1,250 billion barrels (Gb) as ultimate recoverable oil for the globe. Nevertheless, the analysis of the middle section of the curve was not straightforward because few technical databases were available in 1956, and available reserve estimates were reported poorly. Therefore, the curve could not be "filled in" without making an extra assumption about the maximum production rate. Hubbert assumed a maximum production rate 2.5 times greater than the rate in 1956, an assumption that resulted in an extremely low prediction. With those numbers, Hubbert estimated that the peak would occur around the year 2000 (Hubbert 1956, p. 22). Of course, he was not forecasting the aboveground constraints posed by the two oil crises of 1973 and 1979 or the political struggles and armed struggles in the Middle East.

5.3   Hubbert's Curve

By 1962, Hubbert had developed his model in a mathematical fashion, presenting it in *Energy Resources*, a report to the Committee on Natural Resources of the National Academy of Sciences-National Research Council. Hubbert explicitly applied the mathematical derivative of a logistic curve to model two variables: annual rate of discoveries and the annual production rate for oil (Hubbert 1962).

Before Hubbert's application, logistic curves have been used in biology. The classic logistic curve is credited to Verhulst, who used it in 1845 in connection with biological populations. It was used to propose that population growth increases in time to a midpoint (t_m) and then decreases to zero as population stabilizes around a maximum limit. In this application, where there is no negative growth, total population stays constant at a given level (U), rendering a plot that resembles an S-shape (Fig. 5.3). In the 1920s, Pearl and Reed used the logistic curve to model the US population.

In the resource depletion jargon, the derivative of the logistic curve is now known as Hubbert's curve. Since 1962, it has been utilized widely to model annual oil production. There are other similar curves that have been used also, such as the Gauss curve, the Cauchy curve, the sine wave, and even the parabola. These alternatives give similar results for the upper part of the curve, which is the more important part, but the Hubbert curve is the easiest to construct. It is interesting to note that the Hubbert graph for oil production in his famous 1956 paper (Fig. 5.1) has a fatter top than computed with the above formula, probably because he had to draw it using templates or an abacus, so his published graph was only a heuristic approximation of the actual curve.

Since oil needs to be found prior to its extraction, there should be a connection between discoveries and extraction. According to Hubbert's calculations of 1962, proved reserves coming from discoveries in USL48 had peaked in 1956, while the production peak was expected to occur around 1966–1967, a lag of 10.5 years. Nevertheless, Hubbert was using proved reserves, while probable reserves should also be included in this kind of analysis (Laherrère 2005). Nowadays it is accepted that the global discovery cycle peaked in the early 1930s. Furthermore, not all countries are characterized by a single discovery cycle, and there are other constraints to the Hubbert model that need to be understood. In particular, it is to be noted that it is a symmetrical curve, whereas the production curve of an individual field is generally asymmetrical.

Laherrère (2000) has pointed out that a simple Hubbert curve may be applied ideally only in the following cases:

1. Where there are a large number of operators acting independently and randomly in many fields, like in the USA (over 20,000 oil companies). There are several examples of inadequate populations, as Illinois and Ohio with few fields, or Alaska and the North Sea, where a few giant fields came on production simultaneously after the pipeline connections were made.
2. Where exploration follows a natural pattern unimpeded by political events or significant economic factors, as, for example, during the US quotas in the 1960s, the high prices in the 1970s and early 1980s, or when OPEC artificially cut production: Hubbert modeling should not be used for the swing producers as those in the Persian Gulf.
3. Where a single geological domain having a natural distribution of fields is considered; political boundaries should be avoided.

Fig. 5.3 Logistic curve with derivative. Processes represented by logistic curves grow at increasing rates until they reach a maximum rate at time t_m, and slow down their growth while they approach asymptotically to an ultimate limit U. Due to its popularization by King Hubbert, the derivative of the logistic curve is commonly known as "Hubbert's curve" in the jargon of the oil and gas industry

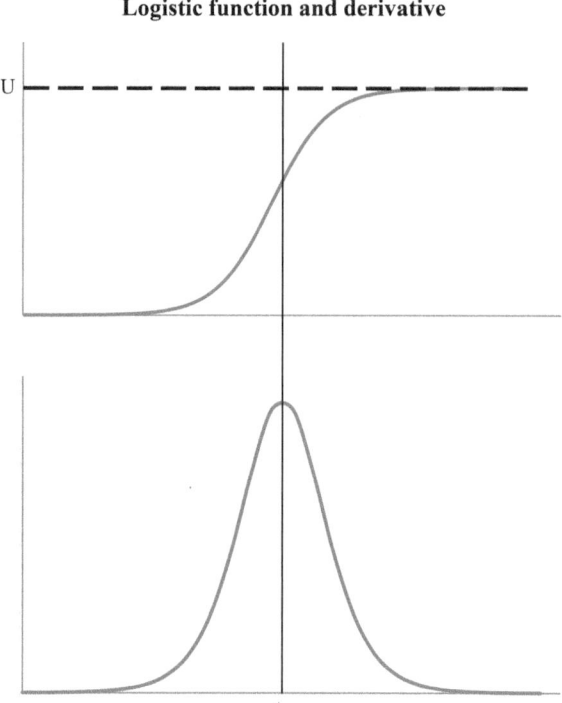

Logistic function and derivative

Hubbert himself did not appreciate these constraints since he worked on the US lower 48 and the world as a whole, prior to significant OPEC intervention. Other techniques are suitable for other specific problems, such as the "plateaus" obtained in the models developed by Jean Laherrère and Jean-Luc Wingert in 2008, which produce better insights of the dynamics of depletion and economic crises (Laherrère and Wingert 2008). Many mathematical and empirical aspects of the Hubbert curve have been explored by Albert Bartlett of the University of Colorado, Adam Brandt of Stanford, and Ibrahim Nashawi and colleagues from Kuwait University (see references). Nevertheless, the contribution of M. King Hubbert, as well as Pogue and Hill and the other pioneers who were warning the world far in advance about the limited nature of oil resources, might be best summarized by Robert Clark, who interviewed Hubbert back in the 1980s: "[Hubbert] makes people, intelligent people who both admire and deplore his opinions, think hard about unpleasant things […] This ability to make people, particularly, the right people, think, will be of inestimable worth. It may be Hubbert's greatest legacy" (Clark 1983). As peak oil occurs in many dozens of oil-producing countries, and as peak oil or something very much like it, seems more and more obvious for the world as a whole, it is very difficult to argue against the statement that he pretty much got it right.

References

Clark RD (1983) King Hubbert. Leading Edge 2:16–24

Hubbert MK (1956) Nuclear energy and the fossil fuels. Shell Development Co., Exploration and Production Research Division, Houston

Hubbert MK, National Research Council (U.S.) (1962) Energy resources. A report to the Committee on Natural Resources of the National Academy of Sciences—National Research Council. National Academy of Sciences-National Research Council, Washington, DC

Laherrère JH (2000) Learn strengths, weaknesses to understand Hubbert curve. Oil Gas 98:63–76

Laherrère JH (2005) Forecasting production from discovery. In: ASPO 4th international workshop on oil and gas depletion, Lisbon. http://www.cge.uevora.pt/aspo2005/abscom/ASPO2005_Laherrere.pdf. Accessed 29 July 2012

Laherrère JH, Wingert J-L (2008) Forecast of liquids production assuming strong economic constraints. In: Seventh annual ASPO conference, Barcelona. http://aspofrance.viabloga.com/files/ASPO7_2008_Laherrere_Wingert.pdf. Accessed 29 July 2012

Murphy DJ, Hall CAS (2011) Energy return on investment, peak oil, and the end of economic growth. Ann NY Acad Sci 1219:52–72. doi:10.1111/j.1749-6632.2010.05940.x

Nashawi IS, Malallah A, Al-Bisharah M (2010) Forecasting world crude oil production using multicyclic hubbert model. Energy Fuel 24:1788–1800. doi:10.1021/ef901240p

Pogue JE, Hill KE (1956) Future growth and financial requirements of the world petroleum industry. Petroleum Department, Chase Manhattan Bank, New York

Bibliography

Section 5.1 is based on
Campbell CJ (2005) Oil crisis. Multi-Science Publishing Co. Ltd., Brentwood (Chapter 4)

Chapter 6
The End of Cheap Oil

In 1998, Colin Campbell and Jean Laherrère published an article in the *Scientific American Magazine* where they estimated the amount of oil that remained to be produced. Campbell got interested in the study of oil depletion around 1969, in Chicago, when he was part of a team making a world evaluation for Amoco (now part of BP). Later, as the manager of the Norwegian branch of the Italian company Fina, he had the company and the Norwegian authorities sponsored a research project on the subject using public reserve data, which later proved to be very unreliable.

These results, published as *The Golden Century of Oil 1950–2050* (Campbell 1991), attracted the interest of Petroconsultants, a company based in Geneva that gathered privileged information from international oil companies to assemble a reliable database on oil activities around the world, including the size of discoveries and drilling statistics (we will return to the history of Petroconsultants in Section 9.4). They invited Campbell to redo the study, but this time using their comprehensive database. Jean Laherrère joined Campbell in the project. The resulting study was published at 50,000 USD a copy but was later suppressed under pressure from a major US oil company. However, Petroconsultants copublished a book, *The Coming Oil Crisis* (Campbell 1997), and agreed that Laherrère and Campbell should accept an invitation to write an article for the *Scientific American*. Thus, "The End of Cheap Oil" was published in March 1998. This paper reawakened scientific and public interest in the precarious position of modern society relative to its necessary oil supplies, a practical and intellectual concept that had lain dormant for more than two decades.

That paper and other related efforts have had considerable impact on reawakening interest in M. King Hubbert's analyses on "peak oil" and its ramifications such that, for example, the Association for the Study of Peak Oil (ASPO) now has 23 national chapters, and four more are in the process of being formed. According to Google Scholar, "The End of Cheap Oil" has more than a thousand recorded citations, ranging from popular science magazines and textbooks to peer-reviewed papers in scientific journals, all of the previous coming from a large diversity of disciplines, not only oil and energy studies but also political science, psychology,

C.A.S. Hall and C.A. Ramírez-Pascualli, *The First Half of the Age of Oil: An Exploration of the Work of Colin Campbell and Jean Laherrère*, SpringerBriefs in Energy, DOI 10.1007/978-1-4614-6064-0_6, © Springer Science+Business Media New York 2013

and cultural anthropology. Fourteen years after its publication, we can say that the article has stood the test of time: it is cited still in numerous works. In this chapter, we present Campbell and Laherrère's (1998) article with some minor modifications in order to make it readable today.

Before moving forward, we would like to recall the importance of defining what we are talking about when we say "oil" or "oil production." Since oil is used mostly as a fuel, the definition of "oil production" usually includes the supplies of *all liquid substances* associated with petroleum or chemical feedstocks, that is, crude oil, extra-heavy and shale oil, natural gas liquids, "refinery gains" (the difference between the volume of total output and the volume of crude oil and other feedstocks that go into refineries), and biocarburants (corn, sugar cane, and cellulose ethanol). This is the definition that Jean Laherrère uses today. Whenever you hear or read any news about "oil," ask which substances are being considered in the definition for there is some confusion with the use and misuse of the term. In 1998, Campbell and Laherrère used the term "conventional oil" as crude oil coming from any source that does not require production technologies significantly different from those used in the mainstream reservoirs exploited at the time. However, experts could never agree on a standard definition of the term, so "conventional oil" has lost its previous significance as extraction technologies have changed in the last decades. Hence, it would be meaningless—and perhaps misleading—to update estimations for conventional oil. The updated data we provide here refers to the most accepted definition today, that is, the supply of all the liquids listed lines above (US EIA 2012). In Chap. 7, we will explore publicly available data collected and corrected by Jean Laherrère.

6.1   Revisiting the End of Cheap Oil

In 1973 and 1979 a pair of sudden price increases rudely awakened the industrial world to its dependence on cheap crude oil. Prices first tripled in response to an Arab embargo and then nearly doubled again when Iran dethroned its Shah, sending the major economies sputtering into recession. Many analysts warned that these crises proved that the world would soon run out of oil. Yet they were wrong. Their dire predictions were emotional and political reactions; for even at the time, oil industry insiders, such as Campbell and Laherrère, knew that they had no scientific basis. Just a few years earlier, oil explorers had discovered enormous new oil provinces on the North Slope of Alaska and below the North Sea off the coast of Europe. By 1973, the world had consumed, according to many experts' best estimates, only about one eighth of its endowment of readily accessible crude oil (i.e., "conventional oil"). The five Middle Eastern members of the Organization of the Petroleum Exporting Countries (OPEC) were able to hike prices not because oil was growing scarce but because they had managed to corner 36% of the market. Later, when demand sagged, and the flow of fresh Alaskan and North Sea oil weakened OPEC's economic stranglehold, prices collapsed.

The oil crunch that we are experiencing now is not so temporary. In 1998, Campbell and Laherrère analyzed the discovery and production of oil fields around the world. Their findings suggested that within the first decade of this century, the supply of "conventional oil" would be unable to keep up with demand. This conclusion contradicts the picture one gets from oil industry and official reports. For example, the US Energy Information Agency (EIA) boasted 1,340 billion barrels (Gb) of oil in "proved" global reserves at the start of 2009. Dividing that figure by the current production rate of about 32 Gb a year as reported by the same agency suggests that crude oil could remain plentiful and cheap for 41 more years (US EIA 2012)—probably longer, because official charts show reserves still growing. It is noteworthy that this last figure, around 40 years of oil to go, has remained the same since 1998.

Unfortunately, this appraisal makes three critical errors. First, it relies on distorted estimates of reserves. A second mistake is to pretend that production will remain constant. Third and most important, conventional wisdom erroneously assumes that the last bucket of oil can be pumped from the ground just as quickly as the barrels of oil extracted from wells in the past. In fact, the rate at which any well—or any country—can produce oil always rises to a maximum and then begins falling gradually back to zero. This is the basic Hubbert analysis that has been used to study oil-producing countries. In some of these countries, oil production exhibits a symmetrical shape with one single maximum resembling the bell curve that M. King Hubbert used to study the USL48, but oil production in many other countries exhibits curves with multiple maxima.

From an economic perspective, when the world runs completely out of oil is not directly relevant: what matters is when production begins to taper off. This is because all of our economic and financial processes are based essentially on an expanding supply of energy, and oil is the most important source of energy for the world. Beyond the tapering point, prices will rise unless demand declines commensurately. This appears to be happening for the world in the first decades of the new millennium: oil prices have climbed to the highest level in history. Many countries are facing severe financial difficulties to repay their debt, not only in the developing world but also among the developed, due in part to high oil prices.

Using different techniques to estimate the reserves of conventional oil and the amount still left to be discovered, Campbell and Laherrère concluded in 1998 that the decline would begin before 2010. According to official figures, global oil production seems to have reached what Jean Laherrère called a "bumpy plateau" around 86 million barrels per day (Mb/d) plus or minus 2 Mb/d since 2005 (Fig. 6.1). This variation of 2 Mb/d, or 2.3% of total production, is less than the difference between the data reported by the US Energy Information Administration (EIA) and the data reported by the International Energy Agency (IEA). Moreover, in 2006, 2007, and 2009 annual production levels stayed below 2005 figures. On a monthly basis, the largest production was reached in January 2012 at 89 Mb/d; between 2005 and 2009, the largest production had occurred in July 2008, right before the Olympic Games in Beijing (Fig. 6.1).

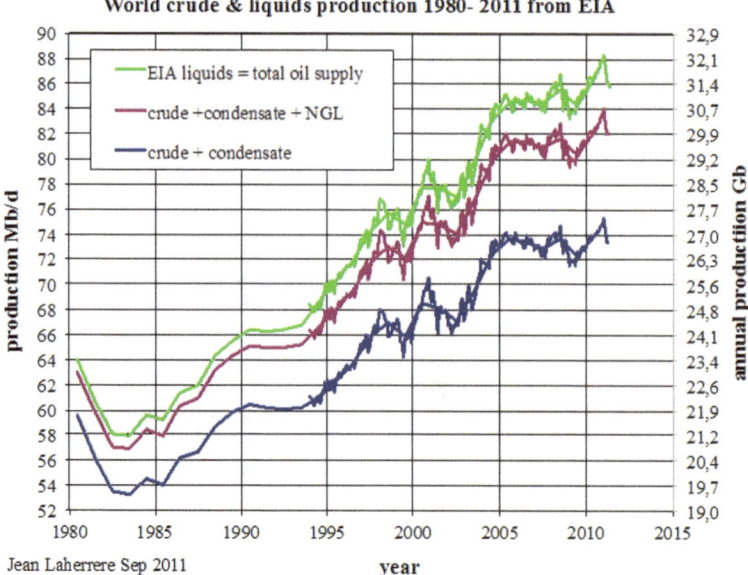

Fig. 6.1 Production of crude oil, condensate, and natural gas liquids (NGL). Crude oil and condensate have remained around 74 Mb/d since 2005; NGL and other liquids have become increasingly important, from less than 6 Mb/d in 1986 to 12 Mb/d in 2011

6.2 Digging for the True Numbers

Campbell and Laherrère spent most of their careers exploring for oil, studying reserve figures, and estimating the amount of oil left to discover, first while employed at major oil companies and later as independent consultants. Over the years, they have come to appreciate that the relevant statistics are far more complicated than they first appear to be.

Consider, for example, the three vital numbers needed to project future oil production. The first is the tally of how much oil has been extracted to date, a figure known as cumulative production. The second is an estimate of reserves—the amount that companies can pump out of known oil fields before having to abandon them. Finally, one must have an educated guess at the quantity of oil that remains to be discovered and exploited. Together they add up to ultimate recovery, the total number of barrels that will have been extracted when production ceases many decades from now (see Sect. 7.3). The obvious way to gather these numbers is to look them up in any of several publications. That approach works well enough for cumulative production statistics because companies meter the oil as it flows from their wells. The record of production is not perfect (e.g., the 2 Gb of Kuwaiti oil wastefully burned by Iraq in 1991, or the oil stolen in many producing countries is usually not included in official statistics), but errors are relatively easy to spot and rectify. The US Geological Survey has estimated that the industry had removed around 1,000 Gb from the earth by the end of 2005, which means some 1,150 Gb after adding production of crude, condensates, and natural gas liquids up to the end of the year 2010.

Getting good estimates of reserves, however, is much harder. Almost all the publicly available statistics are taken from surveys conducted by the entities that publish the *Oil and Gas Journal* and *World Oil*. Each year these trade journals query oil firms and governments around the world. They then publish whatever production and reserve numbers they receive, but they are not able to verify them.

The results, which are often accepted uncritically, contain systematic errors. For one, many of the figures reported are unrealistic. Estimating reserves is not an exact science to begin with, so petroleum engineers assign a probability to their assessments. For example, if, as geologists estimate, there is a 90% chance that the Oseberg field in Norway contains 700 Mb of recoverable oil but only a 10% chance that it contains 2,500 Mb, then the lower figure should be cited as the so-called P90 estimate (P90 for "probability 90%") and the higher as the P10 reserves (Campbell and Laherrère 1998).

In practice, companies and countries are often deliberately vague about the likelihood of the reserves they report, preferring instead to publicize whichever figure, within a P10 to P90 range, best suits their interests. Large estimates can, for instance, help to raise the price of an oil company's stock. On the other hand, sometimes it is advantageous to report lower amounts in order to secure some increases to report in the future, even if no real discoveries are made. Thus, reports are part of the financial strategy of the companies.

The members of OPEC have faced an even greater temptation to inflate their reports because, based on their own internal agreement, the higher their reserves, the more oil they are allowed to export. National companies, which have exclusive oil rights in the main OPEC countries, need not (and do not) release detailed statistics on each field that could be used to verify the country's total reserves. During the late 1980s, 6 of the 11 OPEC nations increased their reserve figures by colossal amounts, ranging from 42% to 197% (Fig. 6.2). The result was the addition of 300 Gb for OPEC members without making any significant discovery. Campbell and Laherrère claimed in 1998 that this increase in reserves was likely to be political instead of geological, and ASPO has insisted the same. The extra amount would boost the production quotas of the countries, allowing them to produce more. It was only in 2007 that Sadad Al-Husseini, former vice president of ARAMCO (who was retired because he wrote a report on peak oil), stated in London that these 300 Gb were indeed speculative resources unlikely to be produced (Al-Husseini 2007).

Previous OPEC estimates, inherited from private companies before national governments took over, had probably been conservative, P90 numbers. So some upward revision was warranted. But no major new discoveries or technological breakthroughs justified the addition of a staggering 287 Gb. That increase is 40% more than all the oil ever discovered in the United States. Non-OPEC countries, of course, are not above fudging their numbers either. For example, in the report of the *Oil and Gas Journal* for 2009, the reserves of 70 nations—including the United States, Russia, China, and India—show no change from 2008 because national agencies did not report changes, even though companies in these countries were extracting

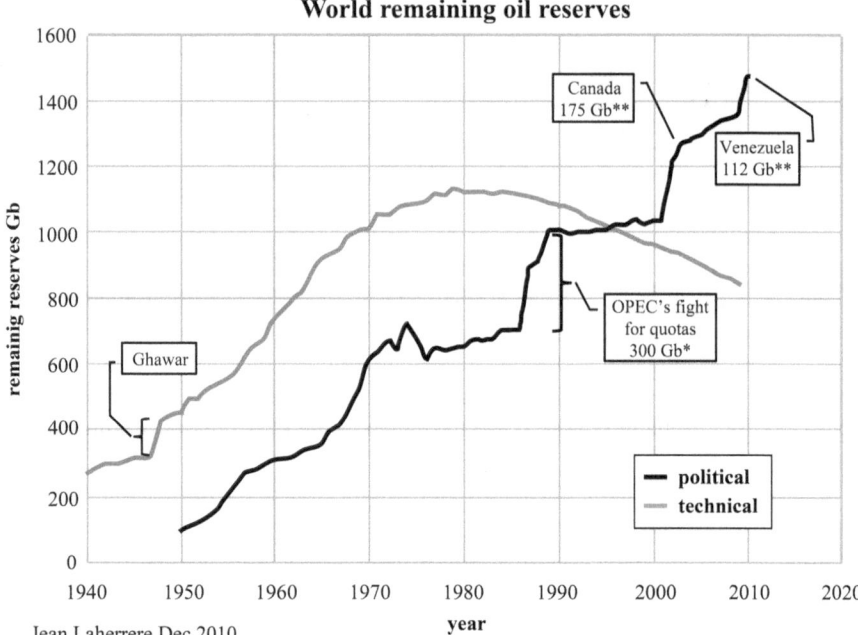

Fig. 6.2 Remaining oil reserves from technical and political sources showing OPEC fight for quotas (*) and the addition of Venezuelan and Canadian tar sands (**). Laherrère defines technical reserves as the addition of proven plus probable reserves (2P) dated back to the original date of discovery without extra-heavy oil; political reserves comprise only proven reserves (1P), include extra-heavy oil, and are not backdated; the initial difference between both curves comes from the omission of probable reserves in political sources and from the incorrect aggregation of reserves. Technical reserves peaked around 1980

oil regularly (US EIA 2012). Because reserves naturally drop as old fields are drained and jump when new fields are discovered, perfectly stable numbers year after year are highly implausible.

6.3   Unproved Reserves

Another source of systematic error in the commonly accepted statistics is the definition of reserves, which varies widely from region to region. In the USA, the Securities and Exchange Commission (SEC) allowed companies to call reserves "proved" only if the oil lies near a producing well and there is "reasonable certainty" that it can be recovered profitably at current oil prices, using existing technology. So a proved reserve estimate in the USA is equal to roughly a P90 estimate. We might consider these to be conservative estimates, as the eventual amount extracted will almost certainly be greater than these numbers. In 2010, the SEC changed the definition of reserves. Now instead of restricting proved reserves to the oil near producing wells, companies can use models to estimate the so-called proved reserves; for reasons of trade secrecy, the companies do not have to disclose precise details about the technology they used to estimate reserve sizes. It is interesting to consider the effect of this change of rules for the shale gas companies in the USA. The big breakthrough in shale gas does not come only from horizontal drilling and hydraulic fracturing technologies but also from the financial rules that allow overestimation of reserves. The goal of promoters becomes not only to produce gas but also to sell part of its interest to major companies such as Exxon, Total, Statoil, and the Chinese CNNOC. The majors, lacking new discoveries to compensate for their production, need "new reserves" on their financial report to prevent the fall of the price of their shares.

Regulators in most other countries do not enforce particular oil-reserve definitions. For many years, the former Soviet countries have routinely released wildly optimistic figures—essentially P10 reserves, which are equal to 3P reserves, that is, the sum of proved, probable, and possible reserves. Yet analysts have often misinterpreted these as estimates of "proved" reserves. *World Oil*-reckoned reserves in the former Soviet Union amounted to 190 Gb in 1996, whereas the *Oil and Gas Journal* put the number at 57 Gb (60 Gb at the end of 2010). This large discrepancy shows just how elastic these numbers can be.

Using only P90 estimates requires additional considerations that are not addressed in many cases. Adding what is 90% likely for each field, as is done in the USA, does not yield what is 90% likely for a country or for the entire planet. On the contrary, summing many P90 reserve estimates always understates the amount of proved oil in a region because the only correct way to total up reserve numbers is to add the mean estimates in each field. The mean estimates can be added because the sum of means yields the mean of the sum, or the total mean; this is not true for P90 estimates because the sum of numbers that occur with a probability 0.9 does not yield another number that will be observed with probability 0.9. For example, if you throw two dices, the probability of a five in each is 1/6, but the probability of getting two fives is 1/36, not 1/6. Moreover, there are several ways to get a ten from the sum of two dices (i.e., a four in the first dice and a six in the second, and vice versa), so a ten occurs with probability 1/12. Adding P90 values is equivalent to adding two fives, a shamefully flawed exercise.

In practice, the median estimate, often called "proved and probable," P50 or 2P reserves, is more widely used and is good enough for a decent estimate. The P50 value is the number of barrels of oil that are likely to come out of a well during its lifetime with probability 0.5—assuming prices remain within a limited range. Errors in P50 estimates tend to cancel one another out, although it is worth noticing that P50 values should not be added, due to the same reasons exposed previously for P10 and P90 values.

In 1998, Campbell and Laherrère were able to work around many of the problems plaguing estimates of conventional reserves by using a large body of statistics maintained by Petroconsultants in Geneva. This information, assembled over 40 years from a myriad of sources, covers some 18,000 oil fields worldwide. It, too, contains some dubious reports, but many errors were detected and corrected.

According to this information, the world had, at the end of 1996, approximately 850 Gb of so-called conventional oil in P50 reserves—substantially less than the 1,019 Gb reported in the *Oil and Gas Journal* and the 1,160 Gb estimated by *World Oil*. The difference was actually greater than it appeared because the value obtained by Campbell and Laherrère represented the amount most likely to come out of known oil fields, whereas the larger numbers were supposedly cautious estimates of proved reserves.

For the purposes of calculating when oil production would crest or peak, the size of ultimate recovery—that is, all the cheap oil there is to be had—is even more critical than the size of the world's reserves. In order to estimate that number, we need to know whether, and how fast, reserves are moving up or down. It is here that the official statistics become dangerously misleading.

6.4   Diminishing Returns

According to most accounts, world oil reserves have marched steadily upwards over the past 30 years (Fig 6.2). Extending that apparent trend into the future, one could easily conclude, as the US Energy Information Administration has, that oil production will continue to rise unhindered for decades to come, increasing almost two thirds by 2020.

As Campbell and Laherrère have explained, such growth is an illusion. About 80% of the oil produced today flows from fields that were found before 1973, and the great majority of these fields are declining. For example, in the 1990s, oil companies discovered an average of 7 Gb a year; in 1997, they extracted more than three times this amount. Yet official figures indicated that proved reserves did not fall by 16 Gb, as one would expect, but rather that they expanded by 11 Gb. One reason is that several dozen governments opted not to report declines in their reserves, perhaps to enhance their political cachet and their ability to obtain loans. A more important cause of the expansion lies in revisions: oil companies corrected earlier estimates of the reserves left in many fields with higher numbers, in particular P90 estimates that by definition were 90% likely to be exceeded. Operators decide to develop a field on the base of net present values (see Sect. 9.1.2) using mean reserves, whose probability is about 40–45%. Shareholders, however, like to have a 90% chance to recover oil, not the 40–45% that operators have to deal with. For financial purposes, such amendments are necessary, but they seriously distort forecasts extrapolated from published reports.

To judge accurately how much oil explorers will discover in the future, one has to backdate every revision to the year in which the field was first discovered—not to the year in which a company or country corrected an earlier estimate. Doing so reveals that global discovery peaked in the early 1960s and has been falling steadily ever since (Fig. 6.3). By extending the trend to zero, we can make a good guess at how much oil the industry will ultimately find.

Campbell and Laherrère used other methods to estimate the ultimate recovery of conventional oil for each country (see Sect. 6.8) and calculated that the oil industry

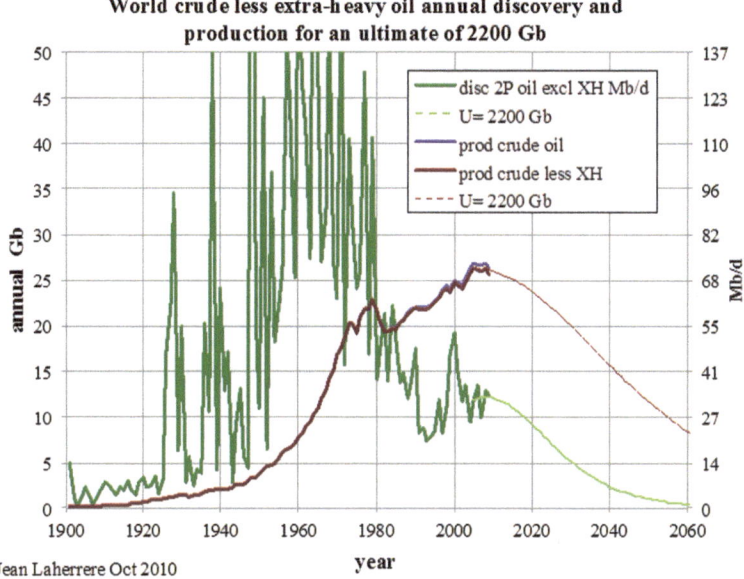

Fig. 6.3 Discovery, production, and projections for an ultimate recoverable (*U*) of 2,200 Gb. At any given year, the area beneath the production curve cannot be greater than the area under the discovery curve. Since the 1960s, discoveries have been decreasing, leaving small room for production to increase

would be able to recover only about another 1150 Gb of "conventional oil." This number, though great, is similar to the lower estimate of the oil that had already been extracted in 2005 (1,050 Gb).

It is important to realize that spending more money on oil exploration will not change this situation necessarily. After the price of crude hit all-time highs in the early 1980s, explorers developed new technology for finding and recovering oil, and they scoured the world for new fields. They found few: the discovery rate continued its decline uninterrupted. There is only so much crude oil in the world, and the industry has found about 90% of the oil lying in fields significantly large to make their exploitation energetically feasible. While there are locations that have not been well explored (e.g., Greenland, ultra-deep water) it is likely that the energy costs of much of this oil (if it is there) would be prohibitive, as the EROI of global oil and gas appears to be declining substantially already (Gagnon et al. 2009).

6.5   Predicting the Inevitable

Predicting when oil production will stop rising is relatively straightforward once one has a good estimate of how much oil there is left to produce; we simply apply a refinement of M. King Hubbert's technique. The global picture is more complicated

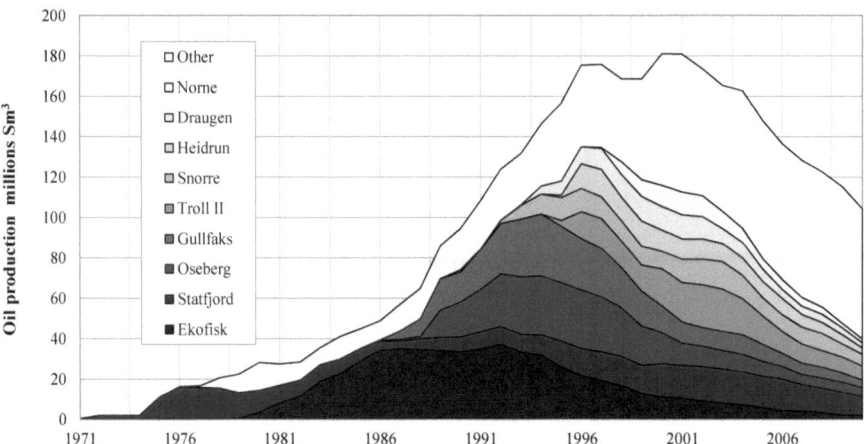

Fig. 6.4 Oil production in Norway showing production of selected fields. Norwegian oil helped to keep oil prices low in the 1990s; the decline of the Norwegian North Sea could not be replaced with other reservoirs after the year 2000

than is the case for an individual field or a nation because the Middle East members of OPEC deliberately reined back their oil exports in the 1970s, while other nations continued producing at full capacity. It is worth mentioning that, since 2002 or so, the world relies principally on Middle East nations, particularly five states near the Persian Gulf (Iran, Iraq, Kuwait, Saudi Arabia, and the United Arab Emirates), to fill in the gap between dwindling supply and growing demand.

The analysis of Campbell and Laherrère predicted that a number of the largest producers, including Norway and the UK, would reach their peaks around the turn of the millennium unless they sharply curtailed production. They did not, and the peak indeed has come to pass (Figs. 6.4 and 6.5). Campbell and Laherrère also had predicted that once 900 Gb had been consumed, production must soon begin to level off or even fall, and this has occurred indeed (Fig. 6.1). World production of oil indeed peaked during the first decade of the twenty-first century, as Campbell and Laherrère—and even Hubbert—had predicted. The situation has been complicated by the global recession and related financial issues, which have greatly decreased demand. So now we bounce along a bumpy plateau, with national economies contracting at about the rate of the oil wells so that "supply and demand" are maintained in approximate balance.

Perhaps surprisingly, the prediction of a peak sometime in the first 10 years of the new millennium does not shift much even if the estimates are a few hundred billion barrels high or low. Craig Bond Hatfield of the University of Toledo, for example, conducted his own analysis in 1997 based on a 1991 estimate by the US Geological Survey of 1,550 Gb remaining (55% higher than the figure of Campbell and Laherrère). Yet he concluded, similar to Campbell and Laherrère, that the world would hit maximum oil production within the 15 years following the year 2000.

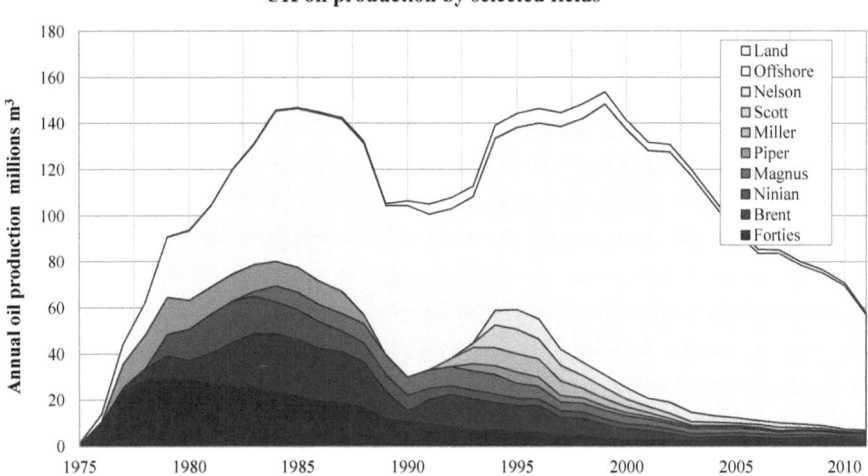

Fig. 6.5 Oil production in the United Kingdom showing production of selected fields. British oil contributed to bring the world out of the oil crises in the 1980s; despite the recovery in the mid-1990s, smaller reservoirs could not compensate for the decline of the British North Sea after the year 2000

John D. Edwards of the University of Colorado published, in 1998, one of the most optimistic estimates of oil remaining—2,036 Gb—although he conceded that the industry has only a 5% chance of attaining that very high goal. Even so, his calculations suggested that "conventional oil" would top out in 2020. At this time what is clear is that oil production, especially that of conventional crude oil, has ceased increasing since about 2005 despite growing prices. This is a remarkable fact and has many implications.

6.6 Smoothing the Peak

Factors other than major economic changes could speed or delay the point at which oil production begins to decline. Three in particular have often led economists and academic geologists to dismiss concerns about future oil production with naive optimism.

First, some argue, huge deposits of oil may lie undetected in far-off corners of the globe. In fact, that is very unlikely. Exploration has pushed the frontiers back so far that only extremely deepwater and polar regions remain to be fully tested, and even their prospects are now reasonably well understood. Advances in geochemistry and geophysics have made it possible to map productive and prospective fields with impressive accuracy. As a result, large tracts can be condemned as barren. Much of the deepwater realm, for example, has been shown to be absolutely non-prospective for geologic reasons.

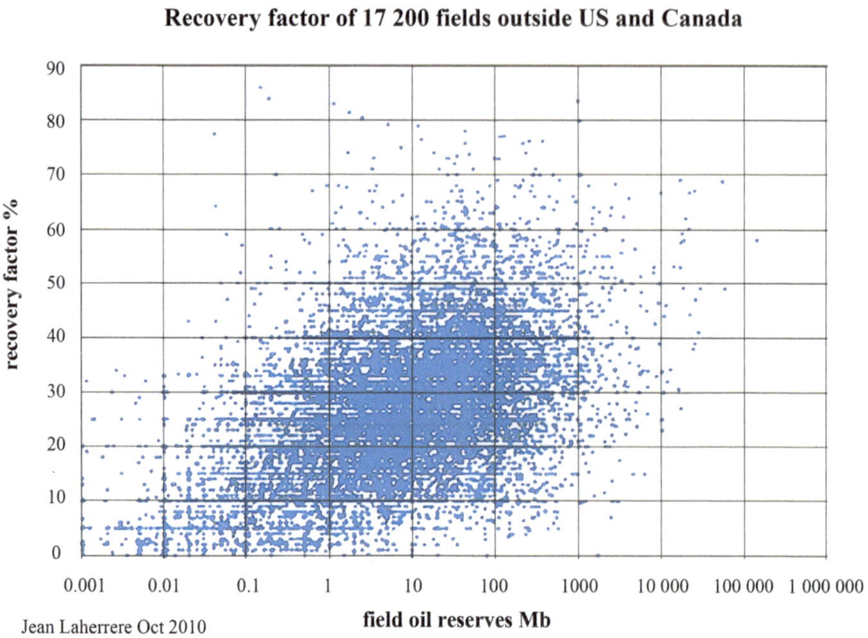

Recovery factor of 17 200 fields outside US and Canada

Jean Laherrere Oct 2010   **field oil reserves Mb**

Fig. 6.6 Recovery factors (RF) and reserves by field; the horizontal axis is in logarithmic scale. RFs below 40% are more common than above 40%; also the RF in the smallest fields tends to be small too. Both variables have great dispersion with no apparent grouping around a single value

A second common rejoinder is that new technologies have steadily increased the fraction of oil that can be recovered from fields in a basin—the so-called recovery factor. In the 1960s oil companies assumed as a rule of thumb that only 30% of the oil in a field was typically recoverable; now they bank on an average of 40% or 50%. That progress will continue and will extend global reserves for many years to come, the argument runs. Recovery factors are often unreliable because the volume of oil in the reservoir is very hard to check even when the field is depleted. The range of recovery is huge, from less than 1% to almost 90%, with good fields being around 50%—as in the North Sea—but no apparent grouping around any value (Fig. 6.6). Hence, neither the mean nor the median is significant.

The dream of boosting recovery factors using technology is unrealistic: a poor field, which has a compact and tight reservoir with less than 3% porosity and about 1–2% recovery factor, cannot be transformed into a porous reservoir with a recovery factor of 50%. Technology allows us to extract oil at faster rates in poor reservoirs with horizontal drilling and hydraulic fracturing, but cannot change the geology of a rock. In other words, we may squeeze some oil drops from a bad reservoir but at a high cost. The ultimate limit for rational exploitation would not be any monetary index, but the amount of oil derived from a reservoir relative to the amount of oil— that is, oil-derived energy and materials—invested in it.

Of course, advanced technologies will buy a bit more time before production starts to fall. But most of the apparent improvement in recovery factors is an artifact of reporting. As oil fields grow old, their owners often deploy newer technology to slow their decline. The fall off also allows engineers to gauge the size of the field more accurately and to correct previous under- or overestimation.

Another reason not to pin too much hope on better recovery is that oil companies routinely count on technological progress when they compute their reserve estimates. In truth, advanced technologies can offer little help in draining the largest basins of oil, those onshore in the Middle East where the oil needs no assistance to gush from the ground.

Last, business analysts like to point out that the world contains enormous caches of "unconventional oil" that can substitute for crude oil as soon as the price rises high enough to make them profitable. There is no question that the reserves are ample: the Orinoco oil belt in Venezuela and the tar sands and shale deposits in Canada and the former Soviet Union contain a vast amount of recoverable hydro-carbons. However, their exploitation is not as profitable as that of regular oil, and even the prospects to develop them support the main thesis: cheap oil is over.

Theoretically, these unconventional oil reserves could quench the world's thirst for liquid fuels as conventional oil passes its prime. But the industry is under hard pressure to get the money needed to ramp up production of unconventional oil quickly enough. An excellent assessment of these and other possibilities for replac-ing oil was undertaken by Hirsch and colleagues in 2005, who concluded that the time to undertake any such transition would be so long that such a replacement needs to start well before the peak even if any replacement is possible and the enor-mous capital investments are made available.

Additionally, most substitutes for crude oil would exact a high environmental price. Tar sands typically are extracted from strip mines. Extracting oil from these sands and shales creates a great deal more air and water pollution than oil. The environmental costs of extracting Canadian tar sands are already restricting its expansion. The Orinoco sludge contains heavy metals and sulfur that must be removed, so governments may exercise their right to restrict these industries from growing as fast as they could. In view of these potential obstacles, Laherrère's esti-mate is that only 500 Gb will be produced from unconventional reserves, a significant amount, to be sure, but not enough to be a game changer.

6.7 On the Downside

Until 2008 global demand for oil was rising at more than 2% a year. Much of the increased demand has been in developing countries. Since 1980, oil consumption is up about 50% in Latin America and 100% in Africa and Asia, according to the Energy Information Administration. In its *2010 International Energy Outlook*, this agency, in the "reference case," forecasts further growth in consumption of liquid fuels in non-OECD Asia, North America, Central and South America, and the Middle East, all of which amount to a 30% increase by 2035 (110 Mb/d) from the levels of 2007. The International Energy Agency made a similar statement for "oil"; in the "Current Policies Scenario" analyzed in the *2010 World Energy Outlook*, the demand for "oil" is more than 107 Mb/d in 2035, an increase of 28% from the 2009 levels (IEA 2010). In our opinion this demand will be too difficult to fill.

The switch from growth to stasis in oil production already has created economic and political tension, such as the economic crisis of 2008 whose effects are still felt throughout the world. Unless alternatives to crude oil quickly prove themselves, the market share of the OPEC states in the Middle East will continue to rise rapidly. These nations' share of the global oil business passed 50% in 2005, way beyond the level reached during the oil price shocks of the 1970s. While there have been many calls for reducing energy dependence in the USA and elsewhere (indeed by the last eight US governments) the fact is that oil and gas provide almost the same percentage of fuel for the USA as they did in 1970, except this time there is more and more oil and gas coming from overseas.

As Campbell and Laherrère forecasted in 1998, the world has seen radical increases in oil prices, changes far beyond what most had thought within the realm of possibility. That alone was sufficient to curb demand, flattening production into the bumpy plateau where we are now, vindicating their previous arguments. Demand fell more than 10% after the 1979 oil shock and took 17 years to recover; it is impossible to guess when it will recover this time, but by now, many Middle Eastern nations are themselves facing their own midpoint of production. We are quite certain that, should economies recover, world oil production will be unable to grow to any significant extent. As the US National Petroleum Council stated in 2007, "the new dynamics may indicate a transition from a demand-driven to a supply-constrained system" (US NPC 2007).

The transition to the post-oil economy need not be traumatic. If advanced methods of producing liquid fuels from natural gas can be made profitable and scaled up quickly, gas could become the next source of transportation fuel. Indeed, it is possible to power automobiles from natural gas, although it can cover only about half the distance that gasoline-powered cars can cover. Safer nuclear power, cheaper renewable energy, and oil conservation programs could all help postpone the impacts of the inevitable decline of oil. A serious problem is time, for countries should have begun planning and investing some years ago. As of 2012, the vigorous initiatives that are required have not appeared either from the governments or from the private sector. At this point, it seems that the burden will be deposited on the civil societies.

The world is not running out of oil—at least not yet. What our societies are facing is the end of the abundant, cheap, and expanding oil supplies on which all industrial nations have come to depend.

6.8 The Methodology Used by Campbell and Laherrère in 1998

In 1998, Campbell and Laherrère combined several techniques to conclude that about 1,000 Gb of conventional oil remained to be produced. First, they extrapolated published production figures for older oil fields that had begun to decline (see Sect. 7.3.3). According to these calculations, the Thistle field off the coast of Britain, for example, would yield about 420 Mb. In May 2011, the cumulative production of

Thistle was 412 Mb; the field is still producing a mix of oil and 96% water. Second, the amount of oil discovered so far in some regions was plotted against the cumulative number of exploratory wells drilled there. Because larger fields tend to be found first—they are simply too large to be missed—the curve rises rapidly and then flattens, eventually reaching a theoretical maximum (see Sect. 7.3.1). Third, Campbell and Laherrère analyzed the distribution of oil-field sizes in the Gulf of Mexico and other provinces. Ranked according to size and then graphed on a logarithmic scale, the fields tend to fall along a parabola that grows predictably over time (Laherrère 2000). Interestingly, galaxies, urban populations, and other natural agglomerations also seem to fall along such parabolas (Laherrère 1996). Finally, the estimates were checked by matching the projections for oil production in large areas, such as the world outside the Persian Gulf region, to the rise and fall of oil discovery in those places decades earlier (Campbell and Laherrère 1998).

References

Al-Husseini SI (2007) Long-term oil supply outlook: constraints on increasing production capacity. Presentation at the oil and money conference, London, 30 October 2007

Campbell CJ (1991) The golden century of oil, 1950–2050: the depletion of a resource. Kluwer Academic, Dordrecht

Campbell CJ (1997) The coming oil crisis. Multi-Science Publishing Co. Ltd., Brentwood

Gagnon N, Hall CAS, Brinker L (2009) A preliminary investigation of energy return on energy investment for global oil and gas production. Energies 2:490–503. doi:10.3390/en20300490

International Energy Agency (2010) World energy outlook 2010. Organisation for Economic Co-operation and Development; Distributed by OECD Publications and Information Center, Paris; Washington, DC

Laherrère JH (1996) Distributions de type fractal parabolique dans la Nature. C R Acad Sci, Sér 2 (Sciences de la terre et des planètes) 322:535–541

Laherrère JH (2000) Distribution of field sizes in a petroleum system: parabolic fractal, lognormal or stretched exponential? Mar Petroleum Geol 17:539–546. doi:10.1016/S0264-8172(00)00009-X

US National Petroleum Council (2007) Facing the hard truths about energy. http://www.npchardtruthsreport.org/download.php. Accessed 13 June 2012

Bibliography

This chapter is primarily based on the following sources:
Campbell CJ, Laherrère JH (1998) The end of cheap oil. Sci Am 278:60–65

Data used to update the article and figures comes from
DECC (2012) UK annual oil production. In: Oil and gas. Department of Energy and Climate Change. https://www.og.decc.gov.uk/pprs/full_production.htm. Accessed 6 Aug 2012

Statistics Norway (2011) Oil and gas, production and reserves. In: Statistisk sentralbyrå—Forside. http://www.ssb.no/ogprodre_en/. Accessed 6 Aug 2012

US Energy Information Administration (2012) International energy statistics. http://www.gov/cfapps/ipdbproject/IEDIndex3.cfm?tid=5&pid=53&aid=1. Accessed 4 Jun 2012

Chapter 7
What Do We Know About "Peak Oil" Today?

We do not inherit the Earth from our fathers, we are borrowing it from our children. [However] we're not borrowing from our children, we're stealing from them –and it's not even considered a crime.

–David R. Brower, 1995 (Let the Mountains Talk, let the Rivers Run)

In this chapter we will analyze data to support the claims about the basic limited nature of the global oil resources that underpin this entire book. Due to the very nature of the task, but also to the prevailing disinformation practices that permeate oil debates, the estimates presented here have a degree of uncertainty, which we acknowledge. However, we believe that current events are proving Campbell, Laherrère, and the "peakists" to be right in general. We leave the discussion about appropriate policies to others.

This evidence has compelled us, and many before, to speak out against the failures of companies and governments in addressing or communicating the problem. According to our arguments, these shortcomings could be threatening the future of communities and entire nations in some cases. We think the problem is serious and that neither the state nor the private sector is reacting in an appropriate way; we must insist, to the risk of being repetitive or pretentious, that awareness about "peak oil" (and EROI) needs an increased level of attention from the media and society in general, at least comparable to that of global climate change. We believe that our role as scientists is to collect and interpret the data to the best of our knowledge and communicate our findings to society in general.

7.1 Technology and Uncertainty in the Oil Industry

The biggest problem anyone faces when trying to assess the future of oil production is uncertainty. It is amazing to us that one of the most important industries for the modern society, having large resources and access to the best possible technologies,

C.A.S. Hall and C.A. Ramírez-Pascualli, *The First Half of the Age of Oil: An Exploration of the Work of Colin Campbell and Jean Laherrère*, SpringerBriefs in Energy, DOI 10.1007/978-1-4614-6064-0_7, © Springer Science+Business Media New York 2013

still relies on unscientific practices coming from the nineteenth century. Jean Laherrère likes to point out the role of technology in the oil industry by quoting the Greek fabulist Aesop, according to whom the tongue is both the best and the worst tool: it is the key to all knowledge and philosophy, the instrument to establish trade and contracts, the means to pronounce eulogies and marriages; but also, the tongue is responsible for all the wickedness in the world, causing the ruin of empires, cities, and relationships; wars and misdeeds are perpetrated only after being discussed, debated, resolved, and communicated, all by words.

The oil industry uses also the best and the worst tools. While the best technology is used in seismic exploration, extracting and logging, the worst technology is often used in defining the units and measurements, in reporting and communicating important issues about oil resources, and also in accelerating the present production to increase current profits to the detriment of future production and future generations. It is embarrassing to see a trillion-dollar industry hiring some of the best engineers all over the world on the one hand, but still following outdated practices and emitting reports that would be unacceptable for undergraduate students. Some of the most salient issues that Jean Laherrère has detected are the following (Laherrère, personal communication):

- Reports are issued with unofficial units different from the universally accepted metric system: e.g. feet..., barrels, and tons.
- Symbols are used to denote different things in the same document: "M" has been used for "thousand," "million," and "metric."
- Assessment of probabilities is incompetent: P90 reserves from oil fields are added to calculate the P90 reserves of countries, and then added again to calculate global reserves. This aggregation underestimates P90 national reserves and the growth of reserves is partly due to this mistake.
- Quantities are reported with irrelevant significant digits: twelve digits are reported when even the first is uncertain.
- Forecasts are done for long periods into the future using insufficient data from the past: estimations for a certain period should report historical past data for a period of about twice the period that is being forecasted.
- Important data is now inaccessible: incredibly, data from USGS was lost or became inaccessible because it was stored in outdated digital mediums that either degraded to a point that they could no longer be read, or a suitable computer or software which could read the data could not be found.

There is a smokescreen of numbers that could and should be avoided. The task is difficult by itself and having to account for all the artificial uncertainty only makes things worse. Governments are not addressing the issue publicly and the industry is running under a business as usual scenario. Meanwhile, oil supplies have not increased significantly since the year 2005 despite enormous increases in the price of oil. That is why we think that some national and international agreements should be implemented in order to guarantee accurate information. Definitions, symbols, and techniques should be standardized on the basis of the best scientific knowledge available.

7.2   Fossil Fuels Have No Apparent Substitute

Many people believe that renewable energies have the capacity to displace fossil fuels and are the solution to an eventual depletion of the latter. That may not be true at all. Renewable energies are not displacing fossil fuels; instead the *increase* in use of oil, gas, or coal in most years is greater than the *total* amount of wind and photovoltaic output, the so-called "new solar energies" (Fig. 7.1). These new energies are just adding to the mix. Before 2008, the new solar was growing more rapidly from a much smaller base; if it is to overtake fossil fuels, it would have to grow much faster from a larger base. As of this writing, growth in all fuels has decreased since the financial crisis of 2008 and the subsequent recession; this decrease has been especially true for the new solar technologies (wind and photovoltaic).

It is not difficult to find very different accounts about whether or not we should be worried about the immediate availability of oil. Many authors (e.g., Ivanhoe, Deffeyes, Hubbert) have indicated that as of 2010 humanity would have burned about half of the oil it will ever burn, everything else remaining equal. On the other hand, other studies (see Sects. 9.2–9.4) suggest that we have not burned a significant amount of the oil we will ever burn. Beyond that, some analysts (e.g., Simon (1998), Lynch 1998, 2001, 2008, 2009; see references) say we will never run out of oil, that the economic process itself will always find more oil and if not, it will provide substitutes. Probably the majority of Americans believe that, essentially, there should

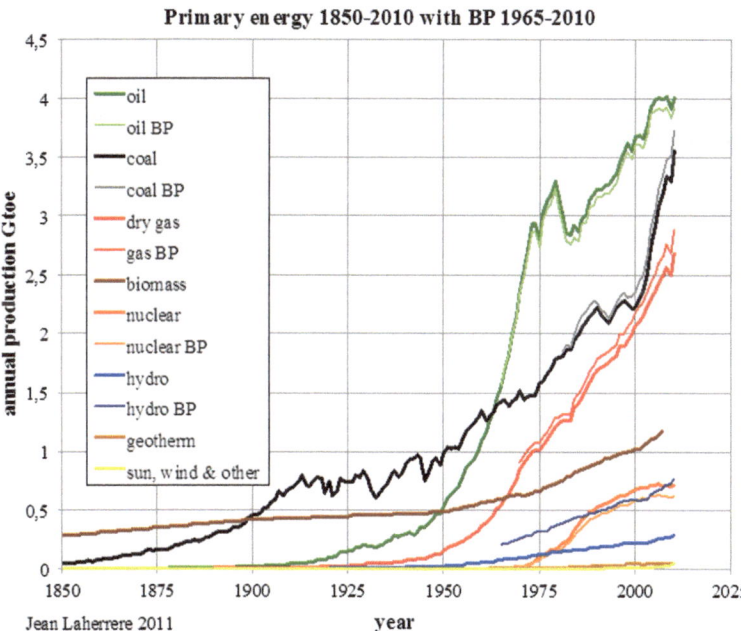

Fig. 7.1 Production of primary energy from different resources in billions of tons of oil equivalent (Gtoe). The difference in hydropower is due to the conversion factor used in BP's statistical review BP (2012)

be no concern about future oil because "scientists and engineers will come up with something." How can we evaluate the veracity of these very different statements? How can we know how much oil is yet to be extracted from the ground? Which assumptions do we have to make in order to produce such figures?

7.3 Ultimate Recoverable, Cumulative Production, and Discoveries

In order to forecast future oil production, we need to estimate the following items:

1. Ultimate recoverable—the volume of oil that can be recovered from worldwide reservoirs at a profitable rate (both in terms of energy and dollars) under the current technology.
2. Past cumulative production—the amount of oil that the industry has pumped out of the ground in the past.
3. Future cumulative production—the difference between the previous two quantities yields the remaining amount of oil that is likely to be extracted in the future.

The total volume of future cumulative production (3) cannot be pumped in one day or in a single year. Therefore, in order to forecast annual production, we also need to estimate the path that oil production is likely to follow year by year—for example, stationary process, exponential growth, exponential decay, logistic pattern, and bumpy plateau—and allocate the future cumulative production according to this path. Hubbert, for example, chose the derivative of a logistic curve (see Sect. 5.3).

Even though the previous calculations seem to be straightforward, keep in mind that the estimation of each of these quantities requires vast amounts of other estimates, each of which have some degree of associated uncertainty. For example, the estimation of the ultimate recoverable requires historical data on discoveries around the world, while past cumulative production involves historical data from extraction rates worldwide.

7.3.1 The Use of Creaming Curves to Estimate the Global Ultimate Recovery of Oil

The oil that we can expect to find and extract in an already exploited region—so-called "mature province"—can be estimated by exploiting the regularities that arise when an experiment—such as finding an oil field—is repeated a large number of times as described by the law of large numbers in statistics. Empirically, the larger fields tend to be discovered earlier, so when a province is mature, the volume brought by new discoveries declines year by year, and future discoveries would almost

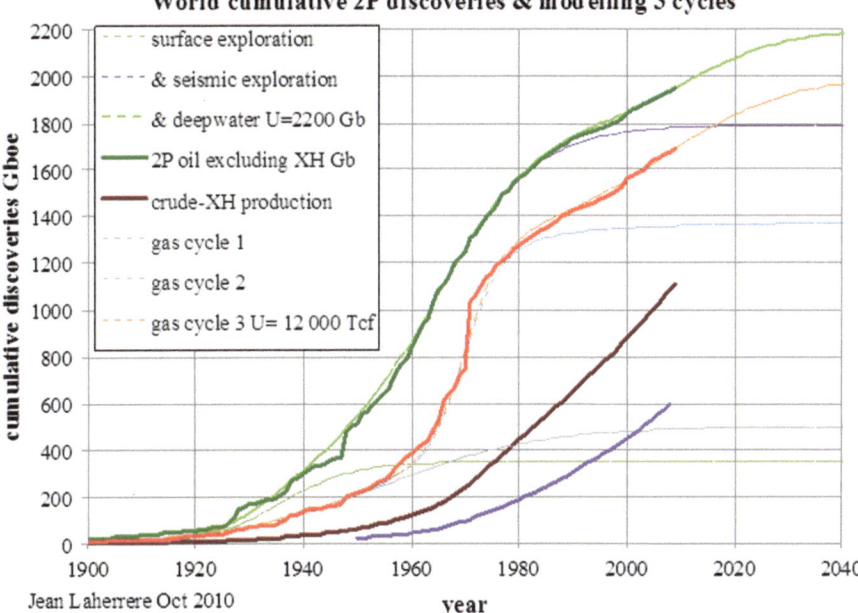

Fig. 7.2 Creaming curve for oil and gas world discoveries under different exploration cycles. Ultimate recoverable seems to be around 2,200 Gb for oil and 2,000 Gb (equivalent to 12,000 Tcf) for gas

certainly be smaller than the ones achieved already unless a groundbreaking technology opens new possibilities for exploration, an event that is becoming less and less likely as the current technologies are already on the edge of our geophysical knowledge. According to a report issued by the US National Petroleum Council in 2007, there are "five core exploration technology areas in which future developments have the potential to significantly impact exploration results over the next 25 years (20 years now). Although the future of these technologies is bright, it is still likely that the trend of decreasing volumes of hydrocarbons discovered with time will continue, although the exploration success rate may continue to improve" (US NPC 2007).

The pattern going from larger to smaller fields can be readily understood through the use of the "creaming curve," a very useful tool designed by Shell in the 1950s for examining the ultimate yield of oil—or natural gas—from any reasonably well-explored region under a given technology. Figures 7.2, 7.3, 7.4, 7.5, 7.6, 7.7, 7.8, and 7.9 show this pattern emerging throughout the globe, most clearly in the Middle East and Europe. The original version of the creaming curve depicted the cumulative discoveries versus the cumulative number of exploratory wells or "wildcats"—an index of exploration effort in the industry. Hyperbolas seem to fit wildcat data nicely; however, information on wildcats is hard to get and old data is not very reliable. On the other side, if we plot time (in years) or the cumulative number of oil fields instead of wildcats, we get the same pattern roughly, though the curve is not very smooth.

Using cumulative discoveries and hyperbolas, Jean Laherrère has model future discoveries and estimated ultimate recoverable several times. In Fig. 7.2, he used three hyperbolas to model discoveries achieved under different exploration cycles: surface surveys (from 1900 to 1950), seismic surveys excluding deepwater (starting in 1930), and deepwater exploration (more than 500 m or 1,600 ft; starting in 1990). It is worth comparing seismic against deepwater exploration: the current deepwater cycle will probably find some extra 150 billion barrels (Gb) in the following 30 years, while the seismic cycle boosted available crude from some 400 Gb found before the year 1950 with surface exploration to 1,700 Gb in the year 1990.

In Figs. 7.2, 7.3, 7.4, 7.5, 7.6, 7.7, 7.8, and 7.9, natural gas discoveries are plotted side by side with oil and condensate (O+C) discoveries. In this context, "condensate" refers to a liquid mix of hydrocarbons that is recovered from natural gas in separation facilities. Since it is a liquid fuel and a substitute for gasoline, Jean Laherrère and many other analysts consider these condensates should be included in the crude oil supply.

Gas reserves are usually measured in trillion cubic feet (Tcf; the T comes from the prefix "tera" which means 10^{12} in the International System of Units). Natural gas has a lower calorific power than oil, so a cubic foot of natural gas has less energy than a cubic foot of oil. The most common equivalence between oil and gas is the following: the energy of one barrel of oil equals the energy of 6,000 cubic feet of gas (in the USA, the exact number is 1 barrel = 5,620 cubic feet); this means 1 Gb (billion barrels of oil) = 6 Tcf (trillion cubic feet of gas). Therefore, in these plots, the volume of natural gas has been divided by six, rendering the energy in gas comparable to that in oil.

It is interesting to notice the amount of new fields that "need" to be found in order to increase oil reserves significantly. The case of the Middle East is illustrative: if the number of fields in the region were to double to 3,000 (a 100% increase), it is not likely for ultimate recoverable in the region to increase more than 5%. The other regions analyzed have not yet reached the flattest part of their curves, but we cannot expect them, by any means, to deliver any volume increase of the size of the discoveries made in the second half of the past century.

A way to check if the global hyperbola provides a reasonable estimation of ultimate recoverable is to add up the estimations provided by the hyperbolas of the regional creaming curves (Figs. 7.3, 7.4, 7.5, 7.6, 7.7, 7.8, and 7.9). The global estimate is 2,200 Gb, while the sum of the different regions outside the USL48 and West Canada yields an ultimate of 2,010 Gb, leaving 200 Gb for USL48 and West Canada. Additionally, there are different estimations for deepwater potential. Jean Laherrère has estimated that the deepwater cycle would yield around 150 Gb, while Colin Campbell has calculated 100 Gb. In any case, the 2,200 Gb seem to be not too far from the mark.

Creaming curves give us an idea about how much oil can ever be extracted, but do not tell us when the oil will be produced, if ever. In other words, creaming curves are not production forecasts. However, we can estimate future production using the information of ultimate recoverable obtained from creaming curves together with the cumulative discoveries and production data.

There are only two places from which we can produce more oil: (1) the oil fields discovered in the past or (2) the oil fields that remain to be discovered in the future. The oil fields discovered in the past can be divided further as follows: (a) depleted, (b) producing oil, or (c) not yet developed. Aggregating the latter two together with

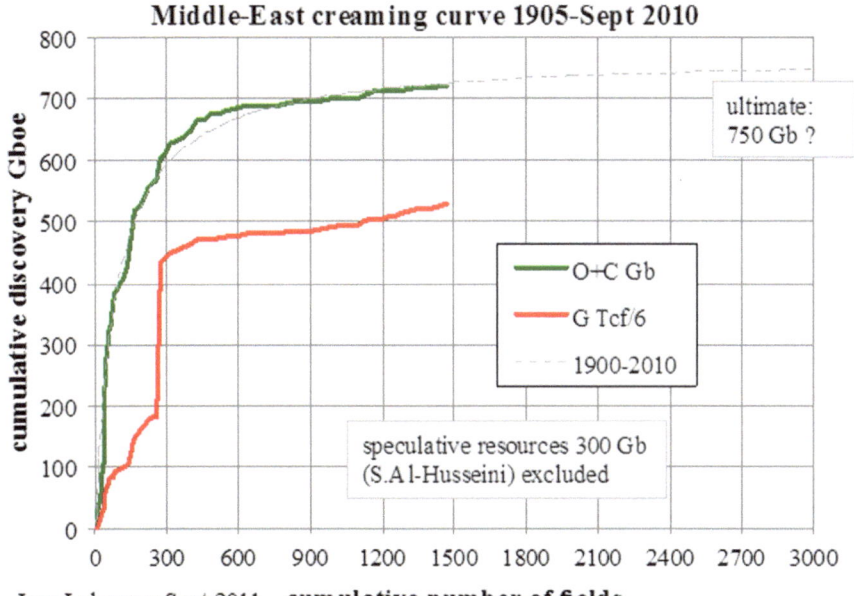

Fig. 7.3 Creaming curve for the Middle East with estimate of ultimate recoverable oil and condensate (O+C) at 750 Gb and excluding the 300 Gb from "political reserves"

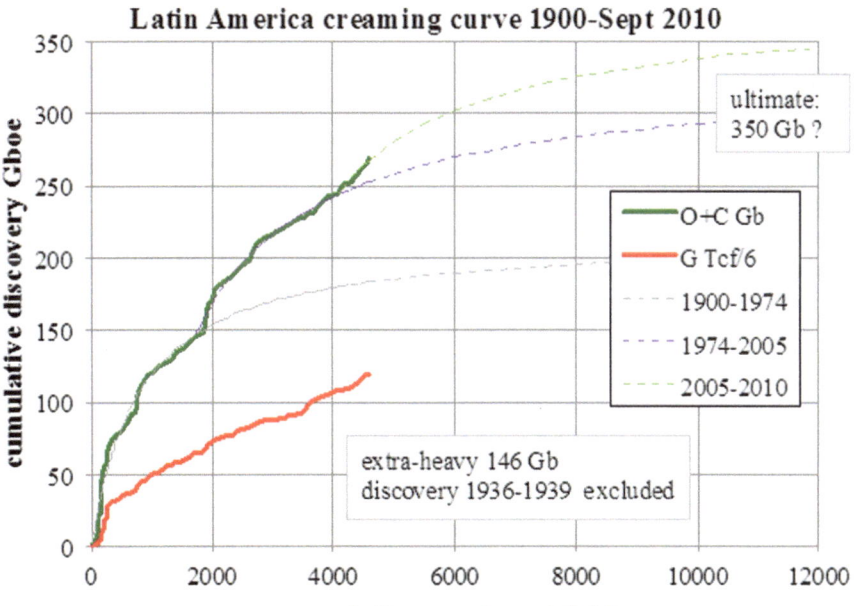

Fig. 7.4 Creaming curve for Latin America with estimate of ultimate recoverable oil and condensate (O+C) at 350 Gb

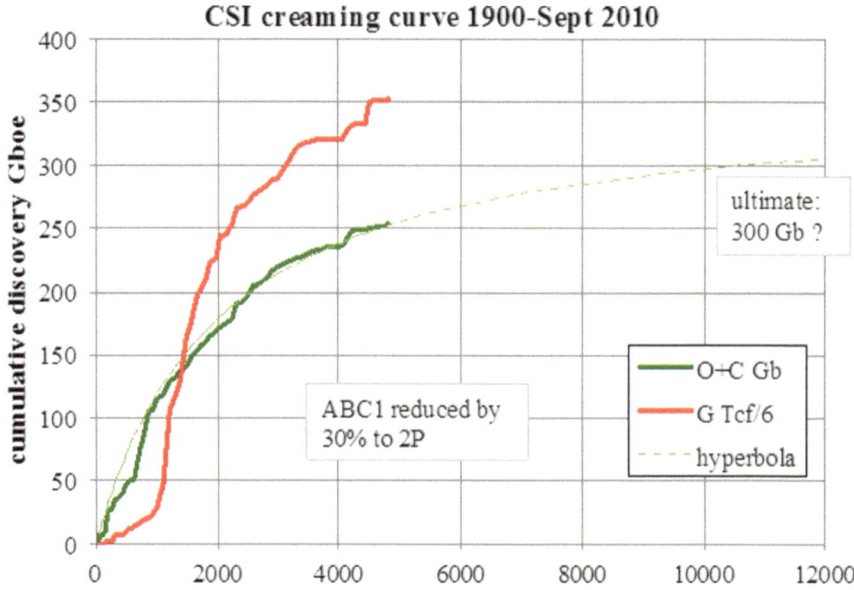

Jean Laherrere Sept 2011   **cumulative number of fields**

Fig. 7.5 Creaming curve for the Commonwealth of Independent States (CIS) of the former Soviet Union with estimate of ultimate recoverable oil and condensate (O+C) at 300 Gb. Not even this historically prolific region rivals the hydrocarbon wealth of the Middle East

future discoveries would yield the ultimate recoverable oil. Since the creaming curves have provided us with an estimation of ultimate recoverable and we can calculate past cumulative production from reports of national agencies around the world, we can estimate the total amount of oil that remains to be produced (Fig. 7.2):

$$\text{Future Cum Prod} = \text{Ultimate Recoverable} - \text{Past Cum Prod} \qquad (7.1)$$

$$\text{Future Cum Prod} = 2{,}200\text{Gb} - 1{,}140\text{Gb} = 1{,}060\text{Gb} \qquad (7.2)$$

7.3.2   Discoveries and Production Cycles: For Oil to Be Extracted, We Need to Discover It

These 1,060 Gb cannot be pumped out of the ground immediately (some of them lay in oil fields that are not yet developed and others in reservoirs that have not been

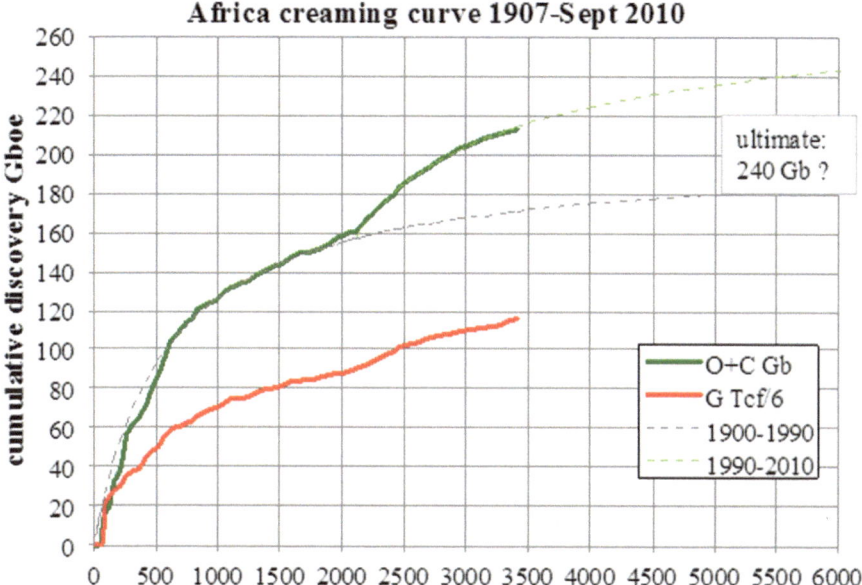

Fig. 7.6 Creaming curve for Africa with estimate of ultimate recoverable oil and condensate (O+C) at 240 Gb

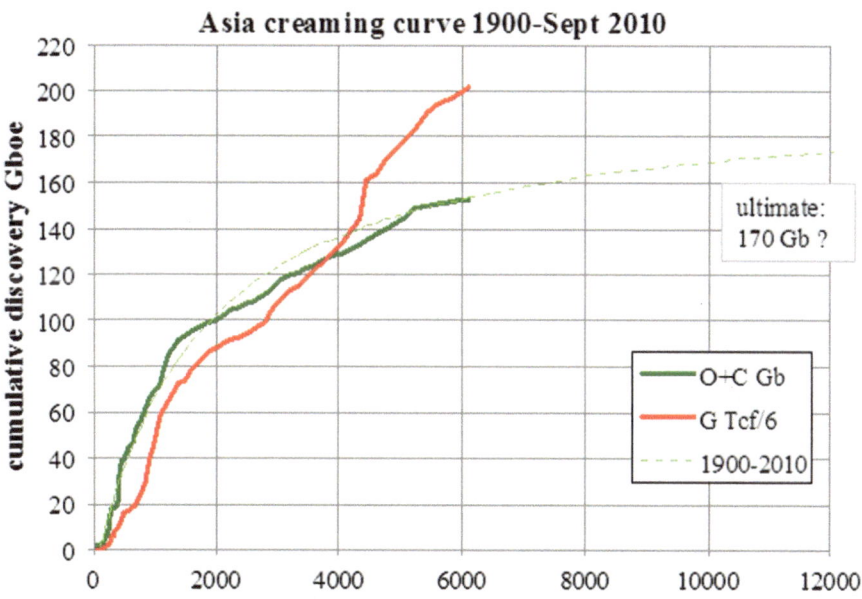

Fig. 7.7 Creaming curve for Asia (except Middle East and CIS) with estimate of ultimate recoverable oil and condensate (O+C) at 170 Gb

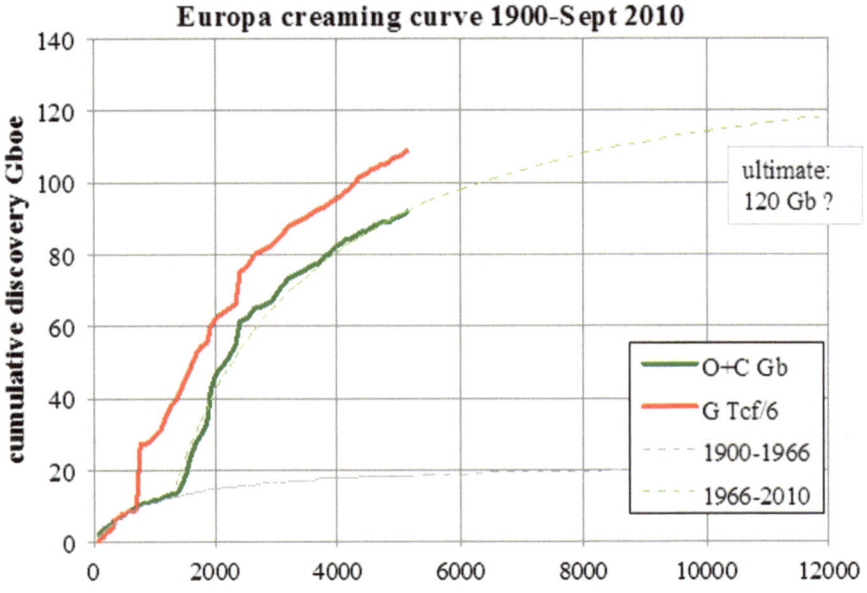

Fig. 7.8 Creaming curve for Europe with estimate of ultimate recoverable oil and condensate (O+C) at 120 Gb

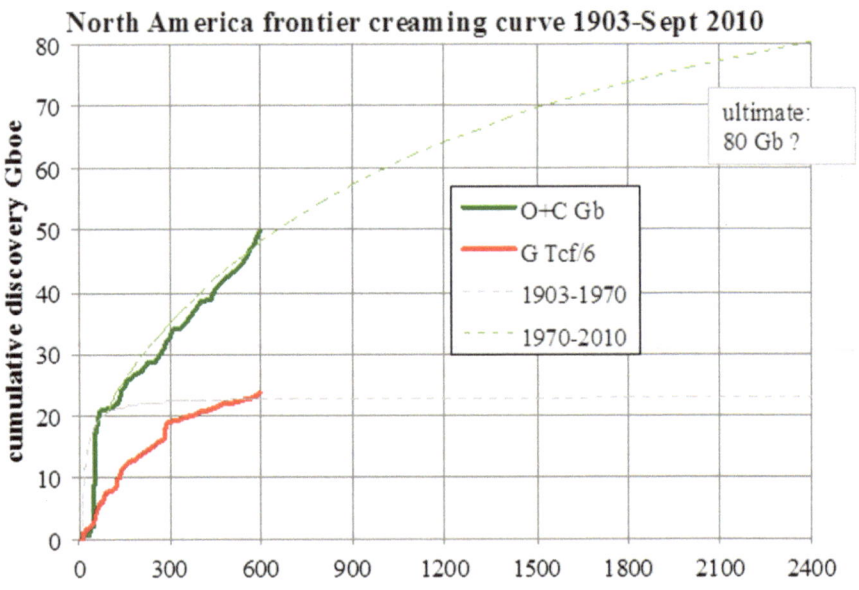

Fig. 7.9 Creaming curve for North America frontier (Gulf of Mexico, Newfoundland coast, Scotian shelf, Alaska) with estimate of ultimate recoverable oil and condensate (O+C) at 80 Gb

discovered); the industry will extract them in the decades to come. What shape will production have in the following years? Is it likely to grow, decline, or stabilize? To answer these questions, there are two pieces of information that we need to consider: (1) the discovery trend in the past and (2) the historical behavior of oil fields.

New discoveries become current reserves or, equivalently, future production, allowing for the uncertainties discussed elsewhere. A large share of today's production is limited by the amount of oil discovered in the recent past. If the discovery trend is increasing now, we need not be concerned about the future for some years, but if discoveries are dropping, we may be worried about a scarcity of oil in the next decades. Moreover, once an oil reservoir is discovered, it needs to be developed; depending on the size and complexity of the project, it may take several years to arrange the legal agreements and build the required infrastructure. So there is a lag between the discovery of an oil field and the point when production begins.

The lag between discoveries and production appears also at the national and global levels. France is one of the best examples to illustrate the relation between discovery and production cycles (Fig. 7.10). The first discovery cycle in France started in the late 1940s and finished in the 1960s, providing the reserves that were exploited during the first production cycle, starting in 1950 and finishing in the late 1970s. Around the same time, the second discovery and production cycles started, but the former peaked and finished earlier than the latter.

While the lags between the discovery and production cycles were different, it is clear that production cannot grow beyond the limit previously imposed by the discovery cycle. In geometrical terms, the area below the production curve in Fig. 7.10—that is, cumulative production—cannot be larger than the area below the discovery curve at any point in time. This is why the oil industry operating in France could not prevent the decline in production during the 1970s; the oil from the previous discovery cycle had been already extracted, so the height of the production had to go down.

Thus, a way to forecast annual production is to look at the discoveries of the past decades. Nevertheless, we must be aware that there are several issues to consider when characterizing discovery trends. For example, discoveries do not follow a smooth path through time; major discoveries occur sporadically and there are many years that are not successful at all. There is, however, a clear decreasing trend in global oil discoveries since the 1980s. In the four decades between 1940 and 1980 the industry made vast discoveries that have become our current production (Fig. 6.3). Despite the growing demand for fossil fuels (Fig. 7.1), and the high prices that have been reached in the last years (Fig. 7.11), the trend of discoveries has been declining during the last 30 years as compared to the previous decades.

In other words, we are extracting more oil than the oil that we are finding. For each barrel produced in the years 2007–2009 less than 0.5 barrels have been discovered. This trend has led to a situation—as in the French case—where past discoveries are not large enough to support either an increase in production or even to maintain the current level for a long time.

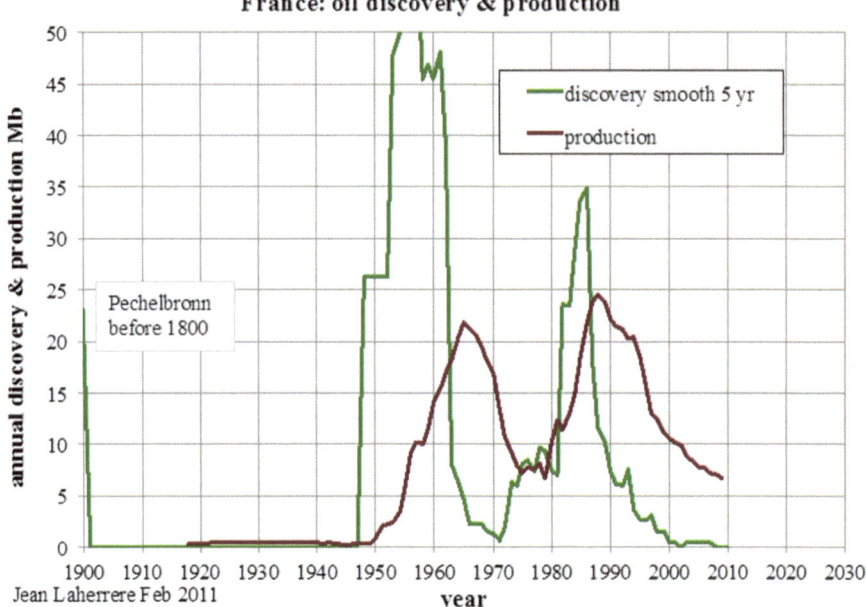

Fig. 7.10 Discovery (5 years average) and annual oil production in France. The two discovery cycles were clearly followed by two corresponding production cycles

7.3.3   Production Data of Active Oil Fields

Estimates for ultimate recovery of currently producing but declining oil fields are straightforward to obtain in general because their production cycle is more advanced and a simple extrapolation will tell you when the field is finished and the quantity of oil you can expect from that field. If an oil field has not reached the decline stage, this extrapolation is not so reliable.

Figures 7.12, 7.13, 7.14, 7.15, 7.16, and 7.17 show the decline of some emblematical oil fields after they enter their decline stage. It is not hard to see that these oil fields have long passed their maximum production rate. Most of them have been placed under enhanced oil recovery (EOR) techniques. From numerous examples we can say that "technology" is useful to accelerate the production cycle allowing companies and governments to pump oil more rapidly than before, but usually without increasing the initial estimation of ultimate recovery in any consistent pattern. In addition, as we stated previously, the effect of technology is already taken into consideration when oil companies calculate their estimates.

As you can see, the great uncertainties around energetic issues can be better understood by examining data. According to Campbell, Laherrère, ASPO, and our own interpretation of the available data, it seems that increasing global oil supply is becoming more and more difficult year after year. Data quality is everything, and with good data you can get a good estimate of how much oil we are likely to produce in the future. International agreements on definitions, symbols, and estimation

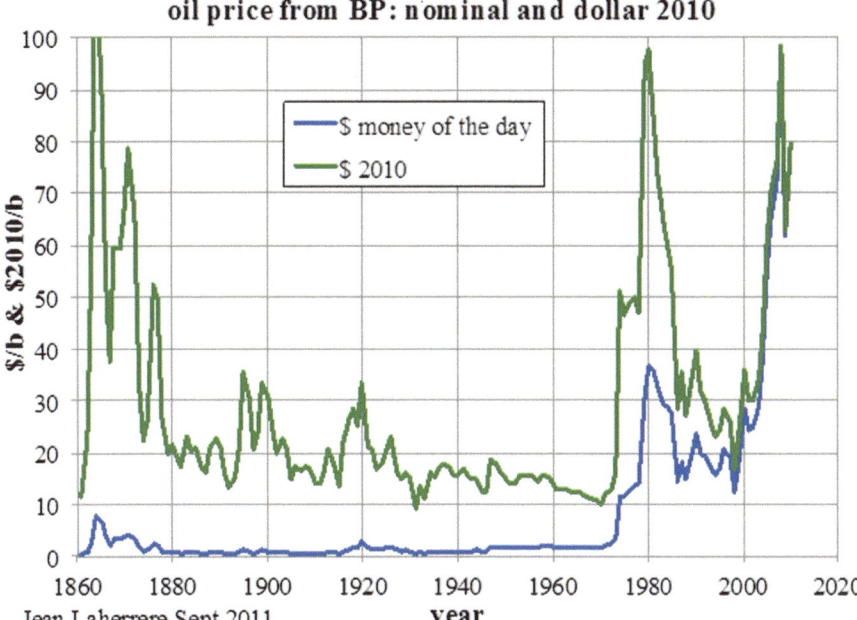

Fig. 7.11 Oil prices since 1860. The recent increases are comparable to those of the oil crises in the 1970s. The low prices in the mid-1980s and 1990s was related to important discoveries like Prudhoe Bay in Alaska, Cantarell in Mexico, or the North Sea

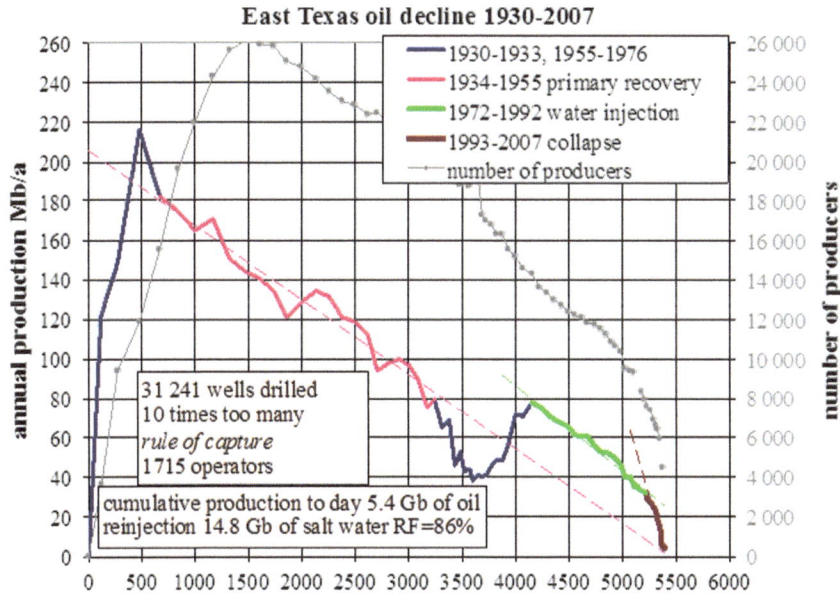

Fig. 7.12 Oil production of the East Texas oil field throughout different epochs. East Texas is famous because of the number of operators allowed to drill due to the "rule of capture" in the USA. Due to overexploitation, this prolific field started declining very soon in 1933

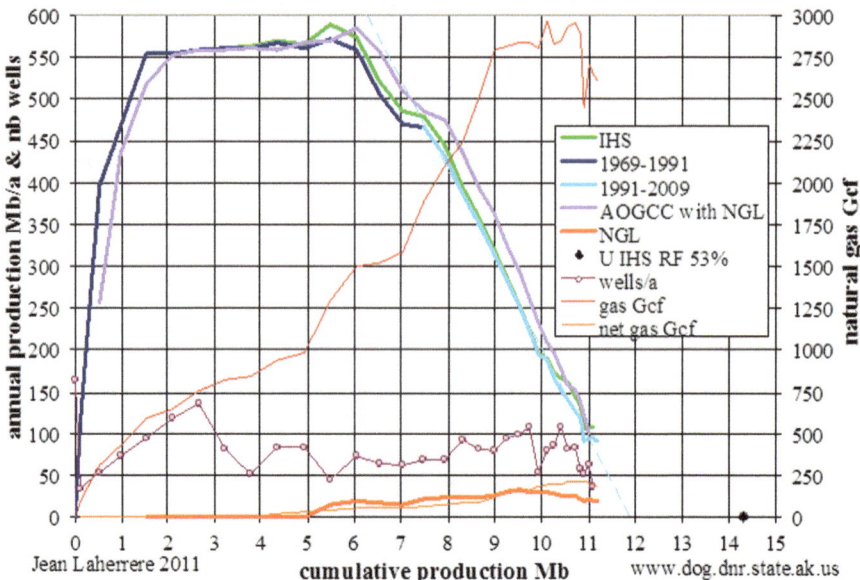

Fig. 7.13 Oil and gas production of Prudhoe Bay, Alaska throughout different epochs. The ultimate recoverable estimated by IHS consultancy at a recovery factor of 53% was too high. Operated by BP, Prudhoe Bay produced a second peak in the national US production, helping to disrupt OPEC's dominance during the 1980s and 1990s

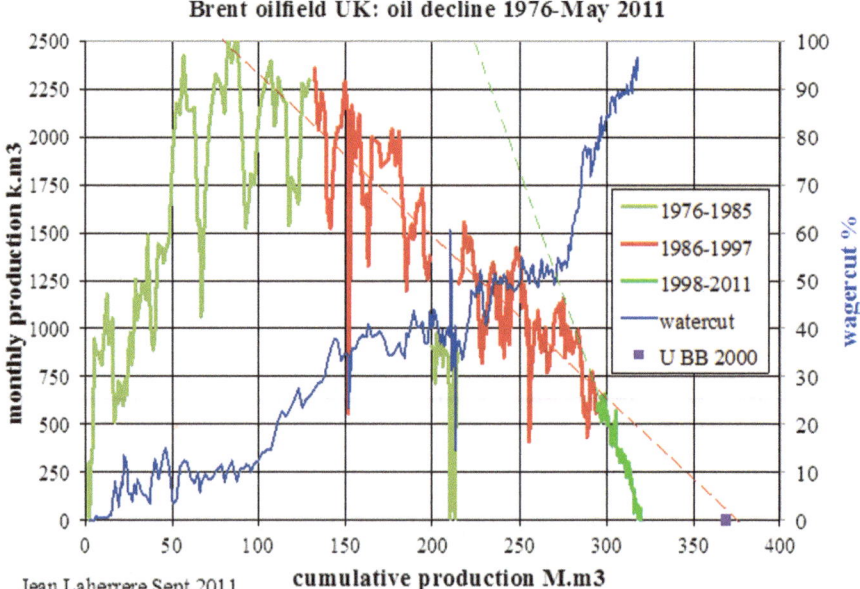

Fig. 7.14 Oil and gas production of the Brent oil field, in the UK North Sea, throughout different periods, showing the ratio of water produced compared to the volume of total liquids produced or watercut. The ultimate recoverable published in the Brown Book (BB) was too high. Once regarded as a standard of quality around the world, Brent oil is now in terminal decline

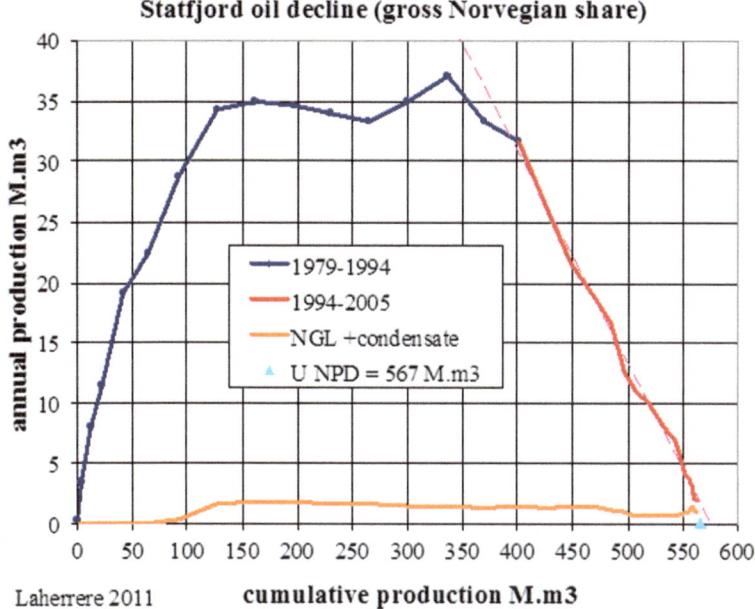

Fig. 7.15 Oil and gas production of Norway's Statfjord oil field, in the North Sea, in two different periods. Statfjord was Norway's largest oil field only after Ekofisk; both fields are now in terminal decline

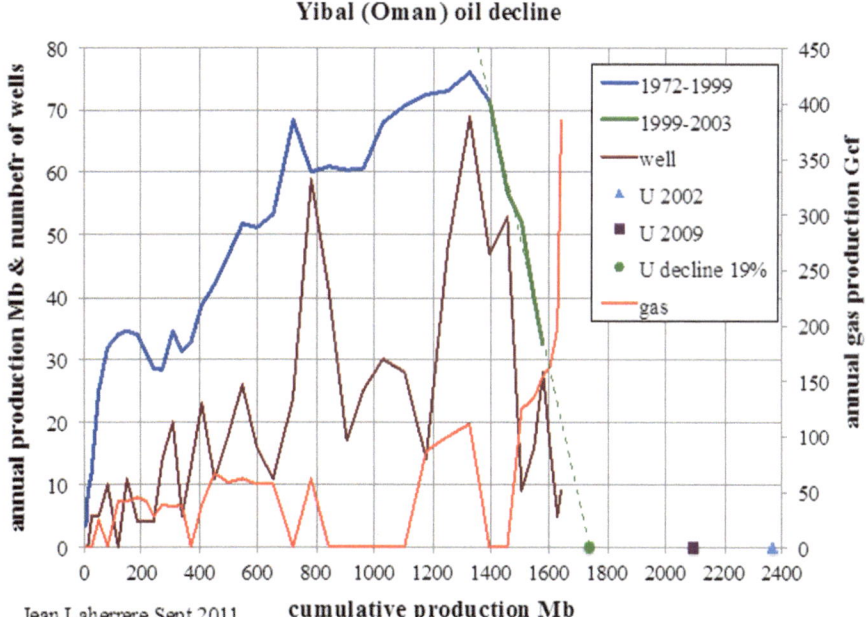

Fig. 7.16 Oil and gas production and number of wells in the Yibal oil field in Oman, with estimates of ultimate recoverable (*U*) calculated in 2002 and 2009. These estimates were too high because extraction was accelerated to increase short-term profits in detriment of future production

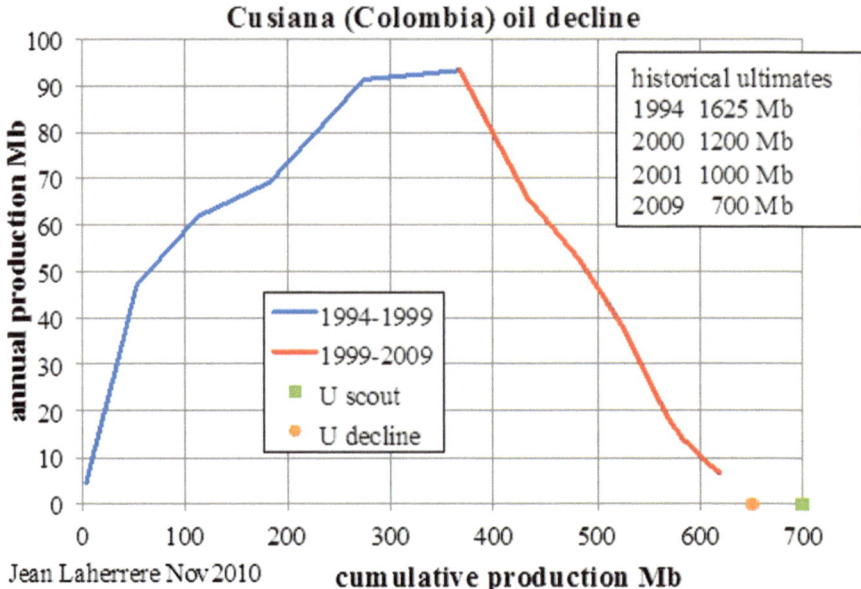

Fig. 7.17 Oil production in the Cusiana oil field in Colombia before and after decline. Cusiana was discovered in a joint venture between total and BP where Jean Laherrère participated. The legend shows BP's overestimations about Cusiana

techniques, as the ones that Jean Laherrère is proposing in a book of this same collection, would help to guarantee more accurate information. Quality information, in turn, would reduce the uncertainty that prevails today, enabling our societies to take better decisions.

References

BP (2012) Statistical review of world energy 2011. BP Global. http://www.bp.com/sectionbody-copy.do?categoryId=7500&contentId=7068481. Accessed 2 Feb 2012

Lynch MC (1998) Crying wolf: warnings about oil supply. In: World oil futures. http://sepwww.stanford.edu/sep/jon/world-oil.dir/lynch/worldoil.html. Accessed 12 Jul 2012

Lynch MC (2001) Closed coffin: ending the debate on "the end of cheap oil." In: World oil futures. http://sepwww.stanford.edu/sep/jon/world-oil.dir/lynch2.html. Accessed 30 Jul 2012

Lynch MC (2008) Letters. Oil Gas J 106:12

Lynch M (2009) "Peak oil" is a waste of energy. The New York Times, 25 August 2009. http://www.nytimes.com/2009/08/25/opinion/25lynch.html. Accessed 6 Jun 2012

Simon JL (1998) The ultimate resource 2. Princeton University Press, Princeton

US National Petroleum Council (2007) Facing the hard truths about energy. http://www.npchardtruthsreport.org/download.php. Accessed 13 June 2012

Bibliography

This chapter is primarily based on the graphs of Jean Laherrère and on the following sources:

Laherrère JH (2005) Forecasting production from discovery. Paper presented at the 4th ASPO international workshop on oil and gas depletion, Lisbon, 19–20 May 2005. http://www.cge. uevora.pt/aspo2005/abscom/ASPO2005_Laherrere.pdf. Accessed 29 July 2012

Laherrère JH (2006) Uncertainty on data and forecasts. Paper presented at the 5th annual ASPO conference, San Rossore, 18–19 July 2006. http://www.oilcrisis.com/laherrere/ASPO2006-JL-long.pdf. Accessed 29 July 2012

Laherrère JH, Wingert J-L (2008) Forecast of liquids production assuming strong economic constraints. Paper presented at the 7th annual ASPO conference, Barcelona. http://aspofrance. viabloga.com/files/ASPO7_2008_Laherrere_Wingert.pdf. Accessed 29 July 2012

Chapter 8
The Formation of ASPO and the Growing Influence of the "Peak Oil" Community

> *In addressing ASPO in Cork, Ireland, I argued that the peakists had won the intellectual argument, except for some minor details about precise timing, but that by and large everyone recognized that there were limits on our capacity to increase the production of crude oil as we have steadily since World War II. [...] But acceptance by knowledgeable people is not enough. The political order should respond.*
>
> –James D. Schlesinger, former US Secretary of Energy

The first question to be asked is why nobody noticed the peak oil issue before? Well, in fact, people did notice. Take Hubbert as one of the first examples. He wrote clearly on this issue and published his analyses in many prestigious and visible locations including the National Academy of Sciences and in published congressional testimony. As we have seen, he was not alone in his views and was followed by others, such as L.F. Ivanhoe who developed his Hubbert Center at the Colorado School of Mines and wrote a quarterly newsletter since 1995 (Ivanhoe 1996), Albert Bartlett who prefers Gaussian curves to "Hubbert curves" (Bartlett 2000), Richard Duncan and Walter Youngquist (Youngquist 1997), and Richard Startzman and his students Al-Jarri and Al-Fattah who plotted oil and gas production of every country using Hubbert curves (Al-Jarri and Startzman 1997; Al-Fattah and Startzman 1999). During the past decade many changes in the production of oil occurred, and many people started to acknowledge the problem. Even the few holdout economists usually acknowledge the issue, although their perception of the timing might be quite different. It is a matter of communicating the message rather than a question of noticing the problem. A new issue is the discovery of methods to develop relatively minor fields, such as the Bakken formation in North Dakota. These developments are causing many people to think that the issue is resolved with new technology. It is not. Bringing Iraq back on line might make a larger difference.

C.A.S. Hall and C.A. Ramírez-Pascualli, *The First Half of the Age of Oil: An Exploration of the Work of Colin Campbell and Jean Laherrère*, SpringerBriefs in Energy, DOI 10.1007/978-1-4614-6064-0_8, © Springer Science+Business Media New York 2013

Part of the explanation relates to the mind-set and working environment of the oil companies. In the 1950s and 1960s, the higher management commonly had an exploration background or could at least call on objective advice. Norman Falcon, the distinguished chief geologist of BP, had a respected place on the Board. Shell of course had Hubbert. More recently financial pressures have called for the appointment of money managers and image makers to senior positions. Even if a geologist is appointed at these levels, the main concerns are not the technicalities of the projects. Take the case of BP's Tony Hayward, who is a geologist with a Ph.D. in geology. Under the new order, if the exploration manager started hinting at the natural limits, as some did, he would be accused of pessimism and a failure to deliver the posture of the dynamic oil finder expected of him. Of course, we are not saying that geologists can tell engineers, financiers, or CEOs how to do their job, but we do believe that many times the exploration departments were effectively relegated to the position of internal contractors, doing what is asked of them. Geologists are supposed to understand nature and the boss is supposed to deal with the economics. If geologists are asked to think first about economics, they will lose their imagination. As Wallace Pratt, former vice president of Standard Oil New Jersey, used to say: "oil is found first in the mind of geologists."

We are not trying to portray an idealized picture of geologists and geophysical sciences; we are just trying to give you an idea about the complexity inside the industry. In fact, many times geologists disagree among themselves about the profitability of an actual project. For example, when Jean Laherrère was in charge of Total's technical services, his company and BP were testing the Cusiana oil field in Colombia. While BP's team recommended stopping the drilling and testing efforts as there was little additional oil to be found, Total advised further tests. The assessments continued and Cusiana turned out to be the third largest of the 37 giant oil fields discovered in the 1990s (Halbouty 2003).

The experience of Colin Campbell can help us to understand how the oil business was changing. In one of its periodic attempts to rebuild a position in Norway, Amoco hired Colin Campbell as a consultant. He arrived in Houston to meet the team and help them prepare the applications to the government for oil concessions. The team was undoubtedly capable in technical terms, but there was a strange lack of direction or sense of judgment. In the meetings, the geologist concerned with each area expounded his interpretation. At the end of one such presentation, Campbell commented that a particular area certainly did not have the resources to justify the expense of development. The geologist, who had developed the report, looked crestfallen and apologized that he had evidently not worked hard enough to develop the prospect. Campbell tried to reassure him, he had done a magnificent job in describing a place that simply lacked the necessary geology. About this episode, Campbell wrote the following:

> I reassured him that he had done a magnificent job in describing a place lacking the necessary geology. His reaction was revealing because it showed that he saw his job not as using his judgment but as applying his skill to employ geological mental gymnastics to make a purse of a sow's ear: if the obvious Upper Jurassic source was not deep enough to generate oil, he would invoke long-range migration, or structural inversion such that what was now

too shallow had previously been deeper. The scope for convoluted hypotheses was limitless. Judgment as such was not part of the job.

Colin Campbell saw things very differently from many of the other geologists. When, as was often the case, he and his colleagues had to propose exploration projects in new areas that did not look promising, the best he could do was hope that commonsense judgment—which often argued against undertaking the drilling—would prove wrong:

> We commonly lacked sufficient information to be absolutely sure, and the only way to know for sure was to drill holes. To get the money to do so from the managerial financiers, we had to pretend that there was a good hope of making money. They themselves risked little, because they could take the cost of failure as a charge against taxes, so that the unconscious taxpayer funded many dry holes. The problem was that they had many alternative opportunities around the world, against which any particular venture had to compete.

Thus, higher management, lacking professional geological qualifications to judge real exploration potential, had been delivered an endless list of similar-sounding prospects for acceptance or rejection based on hypothetical economic and political evaluations that often missed the point. It all involved much theater in the hierarchies of corporate power pyramids and posturing, ending up as little more than exercises in internal or external public relations. It was not so much a case of the blind leading the blind, but rather the blind leading those who had eyes to see but were asked to look the other way. Yet, even in this system, most large valid prospects normally did make it to the top of the pile and delivered easily predictable profitable results, while the cost of the lengthening list of dry holes was happily written off against taxable income.

It is understandable that the economists working in such an environment were misled into thinking that there was no, nor would there be, shortage of exploration opportunity. They in turn conveyed this impression to the investment community, who naturally not only believed what they were told but also had a vested interest in doing so because any talk of decline or limits was anathema to their business.

The financial reporting procedures added to the confusion. Companies were not required to report what they found, but rather only their current "reserves" which led to the much used (and much abused) concept of "reserve replacement." For the financiers, it made no difference if reserves were added by discovery, by acquisition, or by revising upwards what had been underreported. They therefore denied themselves knowledge of the actual trend in discoveries. It was not conspiracy or trickery, but rather a matter of mind-set because the underlying notion of natural limits was simply not there. The accounts were designed simply to describe the current status as if there were infinite opportunities in exploration—like indeed there are for most other businesses. If you want more potatoes, and the price is high enough, the simple solution is to grow more and the system readjusts to deliver a normal economic return. But oil cannot be grown like potatoes—there is only so much.

Consistent with this way of thinking is the widely used parameter of "reserve to production ratio." It states simply that the reserves could support current production for a given number of years with the tacit assumption that more reserves could always be added as the need arose. It absolutely ignores the issue of depletion, which makes the calculation devoid of any sense. It is absurd to imagine that production can be held static for a given number of years and then stop dead, which is implicit in the ratio once the notion of a finite limit is introduced. In short, then, the world approached the end of the last century in denial about the depletion of the resource on which it had come to depend on so heavily. Denial is perhaps too strong a word, as it was not exactly deliberate denial but rather a case of living in the past. *The Coming Oil Crisis* by Campbell was not exactly a best seller when published, but it did begin to contribute to a new awareness. The voices in the wilderness, and there were several of them, began to be heard. A turning point was the article published by Campbell and Laherrère in the *Scientific American* magazine in 1998. In addition to the oilmen themselves, there was the new interest by what might be called the "renewable lobby," promoting solar and wind energy, fuel cells, and even nuclear energy. To that point, they had been primarily motivated by environmental issues, including climate change, but readily saw the significance of the depletion of fossil fuels. At that time there was no focal point for people interested in issues related to peak oil to come together and share ideas and expertise.

8.1 The Formation of ASPO

The Association for the Study of Peak Oil & Gas ("ASPO") is a network of concerned scientists, financial analysts and others in universities and institutions that are committed to study the issue of the peak of world oil and gas supply and evaluate its impact. Its declared mission is the following:

- To evaluate the world's endowment of oil and gas
- To model depletion, taking due account of economics, technology, and politics
- To raise awareness of the serious consequences for humankind

ASPO had its origin in Germany in 2000. Professor Wolfgang Blendinger, an ex-Shell geologist, was the professor of petroleum geology at Clausthal University in Clausthal-Zellerfeld, a small city near Göttingen, Germany. His own experiences in the oil industry gave him an intuitive grasp of depletion. He became interested in the topic and invited Colin Campbell to give a lecture in December 2000 at his university, situated almost literally in the heartland of Germany on the flanks of the Harz Mountains. The lecture was filmed and streamed on the Internet and reached a wide audience.

The German department with responsibility for natural resources, the Bundesanstalt für Geowissenschaften und Rohstoffe (BGR), sent a delegation to the lecture, and over some beers afterwards, Campbell proposed trying to form an organization to formalize the study of depletion. They suggested a meeting with Professor Wellmer, the Director in Hanover, who welcomed the idea but suggested that the best approach would be to keep it informal to avoid inevitable bureaucratic delays. In Norway, Campbell's friends at the Oil Directorate, who had initiated the study of oil depletion 10 years before, joined with enthusiasm on the same informal basis. There was a need to give some identity to this ephemeral grouping, so Campbell started to write a monthly newsletter, at first distributed to a handful of interested people. In his first letter, dated January 2001, he introduced the world to a new term, "peak oil." Little did anybody imagine that the network would grow as it has done, now reaching many thousands of members.

In March 2001, Sarah Astor gave a call to Campbell. Her father-in-law, David Astor, had been the editor of *The Observer* newspaper in England. He had perceptively taken the oil shocks of the 1970s as a very serious matter and was much impressed by a BBC film, *The Last Oil Shock*, to which Colin Campbell had contributed. The Astor family endowed an institute to raise awareness of the issue, which eventually became the Oil Depletion Analysis Centre (ODAC) in London. At first, the center was run by Dr. Roger Bentley from Reading University, who organized a successful workshop at Imperial College in London and began to analyze the data. Jim Meyer later took over the running of the organization to concentrate on raising awareness by distributing news items primarily through the website.

Not long afterwards, Kjell Aleklett, a professor of nuclear physics at Uppsala University in Sweden, paid a visit to Campbell. Aleklett had read the *Scientific American* article and saw the significance of oil depletion in relation to energy policy in Sweden. So, he joined the new organization, which was named the Association for the Study of Peak Oil (the name proved to be successful in communicating the basic concern about a maximum production rate, but also, a bit misleading when interpreted as a prediction of future oil production); Jean Laherrère suggested the inclusion of "natural gas" in the name, so the organization finally came to be known by its current name. By July 2001 interest had grown widely with new members joining the network, such that virtually all European countries were represented by influential scientists in universities and government departments.

The next turning point came in May 2002 when Professor Aleklett organized the First International Workshop on Oil Depletion in Uppsala, to which about 65 people came from around the world. The meeting received wide media coverage. It is not necessary to record all the steps that followed. Subsequent annual workshops were held in Paris, Berlin, Lisbon, San Rossore (near Pisa) in Italy, Cork in Ireland, Barcelona, Denver, and Brussels in 2011. Professor Aleklett, who is the president of ASPO since 2003, has organized a website (http://www.peakoil.net/), as did several other national committees.

Somehow ASPO has become a voice that is heard, although it is nothing more than a loosely sewn network of interested scientists. Prestigious entities including the Deutsche Bank, Aramco, the US Congress, and the International Energy Agency (IEA) have referred to its positions. In parallel with this endeavor, the late Buzz Ivanhoe organized a newsletter in the United States through the Colorado School of Mines, which also began to attract serious attention. Walter Youngquist, now retired Professor of Geology from the University of Oregon, is a supporter of ASPO. In 1997, he wrote a famous book called "Geodestinies: The Inevitable Control of Earth Resources over Nations and Individuals" and a paper with Richard Duncan in 1999, "Encircling the Peak of World Oil Production." Today, members of ASPO are found on 5 continents and more than 30 countries from both the developed and developing world, and more national organizations are in process of formation. Jean Laherrère describes the experience as a "spontaneous generation without any control. Those ASPO nationals are born, grow and should likely die one day."

A growing world awareness of oil depletion and the inevitable peak of production began to spread. The ASPO members and their associates found themselves being invited to an increasing number of conferences around the world. There is no point in listing them all as the list is a long one, but it is worth mentioning some highlights. Since 2000, more than 40 books and about a 150 peer-reviewed articles have been written in relation to the subject. Several films, interviews, and videos have been edited and today are ubiquitous in the Internet. Jens Junghans and Klaus Illum played key roles in organizing a presentation in the Danish Parliament, followed up by a dedicated conference organized by the Danish Society of Engineers. There were the normal spectrum of presentations by geologists, other scientists, members of the financial community, and so on, and they now began to include senior figures from the European Union and government departments.

In London, Roger Bentley and others made an official submission to the House of Lords, followed up later when Chris Skrebowski and Colin Campbell gave a presentation to select committees in the House of Commons in July 2004, as did Charles Hall in 2012. The net began to widen as presentations were given by ASPO members and associates as far afield as Calgary, Houston, Abu Dhabi, India, Australia, Hawaii, and Japan. In California, Kellia Ramares started carrying the story on an Internet news service. In Canada, Julian Darley built up the Post Carbon Institute, with the help of some presentations and a website, addressing primarily the responses to peak oil, but taking peak oil itself as a foundation. He wrote a book in 2004, "High Noon for Natural Gas: The New Energy Crisis," on peak gas with a foreword written by Richard Heinberg, who had already published *The Party's Over: Oil, War and the Fate of Industrial Societies* in 2003. Heinberg is a prolific American journalist and educator who has written extensively on energy, economic, and environmental issues. As senior fellow at the Post Carbon Institute, now located in California (http://www.postcarbon.org/), he has provided important contributions at ASPO meetings.

Professor Kyle Saunders, from the political science department at Colorado State University, and David Summers, mining engineering professor at Missouri University of Science and Technology—then known as University of Missouri-Rolla—started a blog by the name of *The Oil Drum* (also known as TOD in the blogosphere) in 2005. The site contains up to the minute prices of oil and a diverse suite of articles on energy, with accompanying comments and criticisms. It is probably one of the best places to get diverse and generally reliable information about energy. In their first year, they averaged more than 7,000 visits a day and had 2,200 registered accounts (Saunders 2006). Two years later, *The Oil Drum* was rated one of the top five sustainability blogs by Nielsen NetRatings and is currently acknowledged by a diverse collection of public figures, including Congressman Roscoe Bartlett (see Bartlett 2012), Princeton economist Paul Krugman, writer James H. Kunstler, billionaire investor Richard Rainwater, and the English rock band

Radiohead. In 2008, the site received the M. King Hubbert Award for Excellence in Energy Education from ASPO USA (http://www.theoildrum.com/).

A strong supporter of ASPO that deserves special mention was Matthew Simmons. He founded one of the largest and most experienced independent investment banks specializing in the energy industry. In addition to founding Simmons & Company International, he also started the Ocean Energy Institute in Mid-Coast Maine, an organization focused on researching and creating renewable energy sources from all aspects of our oceans. His presentations at several ASPO meetings and his 2005 book *Twilight in the Desert: The Coming Saudi Oil Shock and the World Economy*, questioning OPEC reserves, were a strong push for the propagation of "peak oil" (Simmons 2005). His numerous papers from 1996 to 2010 at his site "Simmons International" (http://www.simmonsco-intl.com/) presented many insightful arguments about the coming oil decline. Unfortunately, Matt Simmons passed away in August, 2010. It was a serious loss to the peak oil community.

Colin Campbell, as one of the founders of ASPO, received special attention from the media. British (three shows on BBC), Dutch, French, Irish, and Korean television crews and a host of independent film producers started to arrive in Ballydehob to film Campbell explaining the essence of the oil depletion argument with the Atlantic breakers below, as a fitting backdrop. Amund Prestegard got the Norwegian television interested in producing a program, but Campbell faced eventual legal conflicts when he declined to change the substance of the message. Maj Wechselmann from Sweden secured support for a film, *Looking for La Luna*, that retraced Campbell's steps from Trinidad to Colombia, where he studied the La Luna formation, the prime oil source rock in northern South America.

Ironically, it was in large measure the invasion of Iraq that prompted this new interest in oil depletion. Many people perceptively saw that the invasion had an oil agenda and began to ask just how important Middle Eastern oil was. The BBC Money Programme went so far as to broadcast a program titled "War for Oil", produced by David Strahan, whom Campbell had already helped with the program called *The Last Oil Shock*. This was broadcast by the BBC too. The doubling of oil prices in the latter half of 2004 really began to concentrate the mind of the public, leading to an avalanche of newspaper articles, including no less than the Wall Street Journal which sent a journalist to Ballydehob for an interview.

The following year, 2005, was full of activity on the oil issue. In February, Robert L. Hirsch, Roger Bezdek, and Robert Wendling published the report "Peaking of World Oil Production: Impacts, Mitigation, and Risk Management," the so-called Hirsch Report (Hirsch et al. 2005), by request of the US Department of Energy (which later discouraged Hirsch from undertaking similar analyses). Later in the same year, Prof. Kjell Aleklett, Dr. Hirsch, and Congressman Roscoe Bartlett gave testimony before for the House of Representatives Subcommittee on Energy and Air Quality on the topic "Understanding the Peak Oil Theory." In October, the Swedish prime minister announced that his government would appoint a commission to make Sweden independent of oil by 2020. ASPO received letters of congratulations for turning Sweden into a new direction. It was also the year when ASPO USA was founded.

In 2006 appeared *The Power of Community: How Cuba Survived Peak Oil*, an award-winning documentary cowritten and coproduced by Faith Morgan, Pat Murphy, and Megan Quinn Bachman, who is now on the Board of Directors of ASPO USA. In May 2003, Faith and Pat attended the second meeting of ASPO and learned that Cuba underwent the loss of over half of its oil imports and survived, after the fall of the Soviet Union in 1990. The documentary film has been translated into seven foreign languages with more than 13,000 copies sold worldwide; an adaptation for the public television series *Natural Heroes* appeared in August 2009.

In 2007, ASPO China was formed. In the same year, Prof. Aleklett was asked by OECD to write a report on the subject that came to be titled "Peak Oil and the Evolving Strategies of Oil Importing and Exporting Countries: Facing the Hard Truth about an Import Decline for the OECD Countries" (Aleklett 2007). A year later, Dr. Euan Mearns, editor of *The Oil Drum*, gave a presentation at the Royal Society of Chemists in Aberdeen, Scotland, about the global energy crisis and its role in the possible collapse of the global economy.

In 2011, the 9th ASPO conference in Belgium held two sessions to advance policy discussions, the first at the Walloon Parliament, in Namur, and the second at the European Parliament, in Brussels. The meeting at the European Parliament focused on transport, energy, and agricultural policy, while the meeting at the Walloon Parliament emphasized the role of regional planning and financial stability in an era of high oil prices. These were the first events where politicians at the Belgian and European level discussed together the impacts of peak oil.

In short, the world has woken up as ASPO and other concerned entities around the world succeed in drawing attention to these critical issues. The evidence is building. The clouds of obfuscation and denial are being swept away. Conferences are being held. Television programs are being made. Governments are being alerted. This awakening is itself a fascinating subject in its own right. Those who only a few years ago were lone voices in the wilderness now find themselves being taken seriously. New opinions, attitudes, and instincts are being formed, although there remain many uncertain points of detail. How successful will the world be in facing the challenges remains to be seen, but at least it becomes increasingly aware of the issue. Some countries may adopt policies to secure oil by military means, which, if successful, would raise the peak and steepen the subsequent decline, making a bad situation worse. Others may begin to find ways to use less and find alternative ways to live. Another option, explored especially by David Murphy in his Ph.D. dissertation (Murphy and Hall 2011), is that any increase in the price of oil above about $80 a barrel would by itself cause economic contraction as more money must be diverted from the rest of the economy to getting the oil (and energy generally). Then the declining economy would lead to reduced oil prices and perhaps another spurt of growth. No one should underestimate the challenges.

8.2   The Rimini Protocol

Perhaps the most promising development of all is the so-called Depletion Protocol. It arose at a conference in London when Colin Campbell was asked to cover not only the problem but offer some ideas for a solution. It did not take long to see that the only real way forward was to cut demand to match world depletion rate. It happened that the then secretary of OPEC, Mr. Lukman, was in the audience, and he came up after the lecture expressing enthusiasm for the idea, which he thought would help reduce the tensions and pressure that OPEC was facing then.

The next step came on April 5, 2003, when Campbell received an invitation by no less than Mr. Gorbachev to attend a conference organized by the Pio Manzu Research Center in Rimini, Italy, entitled "The Economics of the Noble Path." It was a remarkable affair at the Grand Hotel, at which philosophers and thinkers addressed the world condition. Armed police patrolled the corridors, and police helicopters hovered overhead. It seemed a good opportunity to propose again the Depletion Protocol, now renamed the Rimini Protocol for the occasion. It attracted much interest from the Italian press and television. Campbell later drew up the protocol as formally as he could contrive.

8.2.1   The Text of the Depletion Protocol

WHEREAS the passage of history has recorded an increasing pace of change, such that the demand for energy has grown rapidly in parallel with the world population over the past 200 years since the Industrial Revolution;

WHEREAS the energy supply required by the population has come mainly from coal and petroleum, having been formed but rarely in the geological past, such resources being inevitably subject to depletion;

WHEREAS oil provides 90% of transport fuel essential to trade, and plays a critical role in agriculture, needed to feed the expanding population;

WHEREAS oil is unevenly distributed on the Planet for well-understood geological reasons, with much being concentrated in five countries, bordering the Persian Gulf;

WHEREAS all the major productive provinces of the World have been identified with the help of advanced technology and growing geological knowledge, it being now evident that discovery reached a peak in the 1960s, despite technological progress, and a diligent search;

WHEREAS the past peak of discovery inevitably leads to a corresponding peak in production during the first decade of the twenty-first Century, assuming no radical decline in demand;

WHEREAS the onset of the decline of this critical resource affects all aspects of modern life, such having grave political and geopolitical implications;

WHEREAS it is expedient to plan an orderly transition to the new World environment of reduced energy supply, making early provisions to avoid the waste of energy, stimulate the entry of substitute energies, and extend the life of the remaining oil;

WHEREAS it is desirable to meet the challenges so arising in a co-operative and equitable manner, such to address related climate change concerns, economic and financial stability and the threats of conflicts for access to critical resources.

Now it is proposed that:

1. A convention of nations shall be called to consider the issue with a view to an Accord with the following objectives:

   (a) To avoid profiteering from shortage, such that oil prices may remain in reasonable relationship with production cost;
   (b) To allow poor countries to afford their imports;
   (c) To avoid destabilizing financial flows arising from excessive oil prices;
   (d) To encourage consumers to avoid waste;
   (e) To stimulate the development of alternative energies.

2. Such an Accord shall have the following outline provisions:

   (a) No country shall produce oil at above its current Depletion Rate, such being defined as annual production as a percentage of the estimated amount left to produce.
   (b) Each importing country shall reduce its imports to match the current World Depletion Rate, deducting any indigenous production.

3. Detailed provisions shall cover the definition of the several categories of oil, exemptions and qualifications, and the scientific procedures for the estimation of Depletion Rate.

4. The signatory countries shall cooperate in providing information on their reserves, allowing full technical audit, such that the Depletion Rate may be accurately determined.

5. The signatory countries shall have the right to appeal their assessed Depletion Rate in the event of changed circumstances.

We think this protocol certainly deserves urgent attention by the world governments as offering a mechanism for a managed transition to declining oil and gas supply. Demand would be put into better balance with supply, meaning that world prices would be held low, to be in reasonable relation to actual production cost. This would allow the poor countries to afford their needs. Profiteering by particularly the Middle East producers, which in turn leads to massive and destabilizing flows of money, would be avoided. Above all, the consumers would be forced to face the reality of their predicament. Even the Middle East itself would benefit by being forced to prepare by lessening its dependence on oil revenue, which is inevitably set to decline in the future as depletion hits that region too.

There are several issues that need to be tackled for the protocol to work. Jean Laherrère has pointed out that it is absolutely necessary to agree on definitions and measures for concepts such as "production cost" or "amount left to produce." These two points by themselves would take many years to agree upon, although Laherrère is producing another book in this series on definitions in the oil industry.

Interest in the proposal seems to be growing, although not fast enough. A committee of international politicians considered it at the 2005 ASPO Conference in Lisbon. In his book, *The Oil Depletion Protocol: A Plan to Avert Oil Wars, Terrorism, and Economic Collapse*, Richard Heinberg highlights the need for the implementation of the Oil Depletion Protocol and suggests ways in which the protocol can be adopted. The Post Carbon Institute has undertaken the initiative, in association with Heinberg, "to lay the groundwork for and facilitate the successful adoption and implementation of the Protocol." They have created a website that gives the history of the protocol, educational materials, and some actions that people are implementing to reduce their oil dependency by 3% per year" (http://richardheinberg.com/odp).

Speaking of protocols, it is interesting to note the changing reaction of what can be called the Climate Change lobby. We believe that its models of damaging carbon dioxide emissions are flawed to the extent that they are based on extrapolations of oil demand rather than future possible supplies. At first it seemed as if the protagonists were negative to any notion that the natural depletion of fossil fuels would reduce the impact on the environment. But now they seem to become more positive seeing that the Rimini and Kyoto Protocols actually work in parallel, both stressing the importance of restricting demand, albeit for different motives.

In the most recent years the decreased economy of the Eurozone and the cessation of much if any growth of the US economy, probably in response at least in part to the increased oil prices of 2010–2012, and the development of new horizontal drilling/fracking techniques have led to a decrease in the price of oil and to many new articles on a new resurgence of oil production in the United States. While we think these new technologies are important, when you do the numbers, it seems that they cannot, over the next decade, compensate for the decline of the major oil fields that still supply most of the oil in the US. Unfortunately these new developments have led to a cessation of much of the political interest in the peak oil issue. Time will tell how all this plays out, but given the lack of serious preparation for peak oil in the past, it looks even less likely that our governments will take these issues seriously.

References

Aleklett, K (2007). Peak Oil and the Evolving Strategies of Oil Importing and Exporting Countries: Facing the Hard Truth about an Import Decline for the OECD countries (Discussion Paper No. 2007-17). JTRC Discussion Paper Series. Paris, France: International Transport Forum/ OECD. Retrieved from http://www.internationaltransportforum.org/jtrc/DiscussionPapers/ DiscussionPaper17.pdf. Accessed 20 July 2012

Al-Fattah, SM, & Startzman, RA (1999). Analysis of Worldwide Natural Gas Production. Presented at the SPE Eastern Regional Conference and Exhibition, Charleston, WV: Society of Petroleum Engineers. doi:10.2118/57463-MS

Al-Jarri AS, Startzman RA (1997) Worldwide petroleum-liquid supply and demand. J Pet Technol. doi:10.2118/38782-MS

Bartlett A (2000) An analysis of U.S. and world oil production patterns using Hubbert-style curves. Math Geol 32:1–17. doi:10.1023/A:1007587132700

Bartlett R (2012) Peak oil: representative Roscoe Bartlett. http://bartlett.house.gov/issues/issue/default.aspx?IssueID=2057. Accessed 29 Jul 2012

Halbouty, MT (2003). Giant oil and gas fields of the 1990s: An introduction. In: Halbouty MT (ed). Giant oil and gas fields of the decade 1990-1999. AAPG Memoir 78:1–13. Tulsa, OK

Hirsch RL, Bezdek R, Wendling R (2005) Peaking of world oil production: impacts, mitigation, & risk management. US Department of Energy, National Energy Technology Laboratory

Ivanhoe LF (1996) Updated Hubbert curves analyze world oil supply. World Oil 217:91–94

Murphy DJ, Hall CAS (2011) Adjusting the economy to the new energy realities of the second half of the age of oil. Ecol Model 223:67–71. doi:10.1016/j.ecolmodel.2011.06.022

Saunders K (2006) The Oil Drum celebrates its first year today. In: The Oil Drum. http://www.theoildrum.com/story/2006/3/22/0238/50619. Accessed 29 Jul 2012

Simmons MR (2005) Twilight in the desert: the coming Saudi oil shock and the world economy. Wiley, Hoboken

Youngquist WL (1997) GeoDestinies: the inevitable control of Earth resources over nations and individuals. National Book Co, Portland

Bibliography

This chapter is primarily based on the following sources:

Campbell CJ (2005) Oil crisis. Multi-Science Publishing Co. Ltd, Brentwood (Chapter 14)

Campbell CJ (2006) The Rimini Protocol, an oil depletion protocol: heading off economic chaos and political conflict during the second half of the age of oil. Energy Policy 34:1319–1325. doi:10.1016/j.enpol.2006.02.005

Chapter 9
The Other Side

Most of the arguments to belittle the importance of resource constraints in general, and "peak oil" in particular, have come from the oil companies themselves, some official institutions and economic studies on natural resources. On theoretical grounds, conventional economic theories emphasize two main aspects concerning natural resources: the capacity of the industry to find new reserves and to develop substitutes for costly resources. Both argue against the importance of depletion. While we think both arguments are possible in theory, we believe that their development remains seriously incomplete, especially from an empirical perspective. On the other hand, the empirical studies carried out by many reputable agencies—such as the United States Geological Survey (USGS)—seem to have overlooked important issues about the reliability of the data sources they used. In any case, it is very important to provide the data used to support one's perspective. For example, assessments which are based on access to expensive and presumably reliable private data—such as the ones presented by IHS CERA—cannot be verified, and their forecasts have had a dubious record so far. Of course, "The End of Cheap Oil" relied on private data too, but its claims seem to be more realistic when confronted with actual production and prices.

Campbell, Laherrère, and ASPO have sustained a large debate against all these arguments. Throughout the years, they have refined their claims to take into account other points of view. Yet, they remained skeptical about theories and scenarios that project an ever-growing supply for different reasons. Inside the private oil companies, economic reports are aimed at evaluating the profitability of individual projects or the profits of the company at large rather than assessing energy security. National oil companies usually have to fulfill the production or monetary objectives of their national governments. Their role is not to supply the global market but to guarantee the energy supply in their home countries and maximize the financial benefit from their production.

Private consultancies rely on private data provided by the oil companies worldwide. Compared to public data, their information is more accurate, but the results of their assessments cannot be verified and do not seem to match recent developments.

C.A.S. Hall and C.A. Ramírez-Pascualli, *The First Half of the Age of Oil: An Exploration of the Work of Colin Campbell and Jean Laherrère*, SpringerBriefs in Energy, DOI 10.1007/978-1-4614-6064-0_9, © Springer Science+Business Media New York 2013

Furthermore, they are not exempted from external pressures: the companies, their main clients, may want to portray themselves as members of a healthy and strong industry basically to stimulate their investors, impress potential competitors from other industries, keep good relations with friendly governments, or keep the unfriendly ones away from their realm. Finally since these consultancies are in the business to secure a profit, their products tend to be expensive. A few exceptions, such as BP, make their generalized data publicly available.

In the face of this situation, the governments of the developed world have funded agencies that monitor the oil and energy industries to guarantee energy security. However, their situation has been similar to that of the national oil companies in that they seem to be influenced to some degree by the governments that are funding them. Furthermore, they must rely on the data provided by the oil companies worldwide, who may not have the right incentives to cooperate with them.

Borrowing from original writings of Campbell (2005), we present a brief overlook of the economic tools and the internal dynamics inside the oil companies (Sect. 9.1). Then, drawing from the outstanding dedication and patience of Laherrère (2000), we address the *World Petroleum Assessment* published in 2000 by the USGS and the press releases of the consultancy IHS CERA. Due to lack of space, we cannot cover other issues, such as the position of the US EIA or OPEC. Finally Sects. 9.3 and 9.4 are based on our own research.

9.1   Economic Assessments Inside the Oil Industry

At the heart of conventional economic thinking are the well-known ideas that supply and demand determine the production and distribution of goods and services. If wheat prices rise, farmers plant more in the next sowing, natural gas flows a bit more towards the production of fertilizer, steel goes to the production of agromachinery, etc. and the whole system readjusts; it is assumed that farmers can control the complex production process completely, simply by their purchases. In essence, economic theory is built around human agency, which indeed reflects many aspects of economic life but overlooks the natural processes that lay outside human will, treating them as "risks" that need to be minimized. For example, the cost of coal is deemed to be nothing more than the cost of the miners and the capital investment weighed by the perceived risks: the resource itself being there for free. If the reserves were infinitely large, perhaps there would be no need to consider them otherwise than as a gift from nature that can be used to produce more and more, but there are some warning signals in this proposition. In this sense, the ideas and tools used in conventional economic assessments are an expression of the old conception, reinforced in the Bible ("dominion over nature," "go forth and multiply") and other religious texts that have been interpreted as depicting humans as the absolute masters of nature.

9.1.1   Risk

Much of the practical work of economists in the upstream sector of the oil industry is concerned with the management of risk. It is thought that there are recognizable economic trends and that certain economic tools can improve the judgment of oil-men in making business decisions. The industry likes to depict itself as having to face exceptionally high risks, for example:

- Natural risk—weather, the 100-year wave, etc. obstruct their activities.
- Environmental risk—they spill some oil and have to clean it up.
- Exploration risk—they may be looking in the wrong place.
- Geological risk—the geological interpretation may be wrong.
- Development risk—the engineers got it wrong.
- Contract risk—the lawyers did a bad job.
- Labor risk—the workers strike.
- Government risk—new governments may not be friendly to the companies.
- Tax risk—the taxes change, even retroactively.
- Political risk—war, sequestration.
- Terrorist risk—somebody blows it up.
- Corporate risk—their stock suffers, or they are subject to a takeover bid.
- Commercial risk—(discounted) prices fall or (discounted) costs rise (see Sect. 9.1.2).

However, most industries and businesses work with even lower profit margins and higher risks: an arbitrary change in government policy-cutting subsidies can bankrupt the farmer after years of work; the arrival of a supermarket puts long established and successful small traders out of business; the lifting of trade barriers may destroy a local enterprise that was effective in providing both goods and employment. According to Campbell, what distinguishes the oil industry is not the risks it faces but the huge sums involved. While the tax rates can be very high, the profits are even larger. The latter amply cushions most of the risk to which they are exposed.

9.1.2   Discounted Cash Flow

The primary economic challenge in exploration is to model actual or anticipated cash flow. Table 9.1 shows a study of a hypothetical development project, undertaken to see if an exploration drilling would be viable if successful. The parameters are quite simple: gross revenue is production (in million barrels) times oil price (in dollars per barrel). Net cash flow is gross revenue less expenditure (capital and operational) and tax (all monetary values in millions of dollars). If it is positive, there is a profit; if negative, a loss.

Table 9.1 Economic evaluation of a hypothetical field using 10% and 15% discount rates

Year	Production (Mb)	Oil price ($/b)	Gross revenue (M$)	Operative expenditure	Operative income	Capital expenditure	Tax	Net cash flow	Net present value (10%)	Net present value (15%)
0	0.00	20.00	0.00	0.00	0.00	95.00	0.00	-95.00	-95.00	-95.00
1	4.60	21.00	96.60	40.00	56.60	0.00	0.00	56.60	51.45	49.22
2	4.20	19.50	81.90	40.00	41.90	0.00	10.40	31.50	26.03	23.82
3	3.90	18.00	70.20	40.00	30.20	0.00	9.40	20.80	15.63	13.68
4	3.50	16.00	56.00	40.00	16.00	5.00	4.00	7.00	4.78	4.00
5	3.10	16.00	49.60	35.00	14.60	0.00	4.50	10.10	6.27	5.02
6	2.60	15.00	39.00	33.00	6.00	0.00	1.90	4.10	2.31	1.77
7	1.90	16.00	30.40	31.00	-0.60	3.00	-0.20	-3.40	-1.74	-1.28
8	0.00	17.00	0.00	0.00	0.00	7.00	-1.60	-5.40	-2.52	-1.77
Total	23.50		423.70	259.00	164.70	110.00	28.40	26.30	7.22	-0.53

At 10% the project is profitable ($7.22 million); at 15% it is not ($-0.53 million)

Actual oil fields have larger life cycles; the regular life for an offshore field is 25 years

Capital expenditure or investment is the cost of the facilities, including the drilling; operating costs are the running costs of labor, insurance, tariffs on pipelines, and contracted services. Finally, tax is the total amount paid to the government, including royalties.

The next step is to calculate what is called discounted cash flow to determine the present value of future earnings. Thus $1,000,000 that you will receive 5 years from now, at a 10% discount rate, is worth today $1,000,000/[(1+0.1)^5] = $620,921.3.

The sum of each future year's discounted cash flow over the life of the field gives the "present value" (PV). In Table 9.1 we discounted the cash flow at two different rates: 10% and 15%. The higher the discount rate, the less value is assigned to future dollars (i.e., future oil production), or conversely, the more weight is placed in present dollars. Due to the large capital expenditures that have to be done before production begins—that is, in year zero—our hypothetical field is profitable at a 10% discount rate, but not at a rate of 15%. Which discount rate to use is subject to complex and rather arbitrary financial considerations.

9.1.3 Oil Prices and Other Considerations

From this information, companies can also calculate other indicators, such as the "payout," that is, how long to wait until the investment is recouped and the project moves into profit, or the rate of return. Companies normally have what is called a "hurdle rate of return," namely the minimum return that they can accept under their investment policy. These calculations describe the simplest outline of the procedure. There is great scope to make it ever more complex, by addressing multiple scenarios and risking each element using probability theory and so forth. The whole process seems fairly correct at first glance. The geologist provides his or her estimate of reserves; the engineers feed in information about the numbers of wells and producing rates; the construction people estimate how much will cost to build the thing; and a committee of economists is dragged out to pronounce on future oil prices. The calculator whirrs, and out comes the answer: the project flies or it does not.

If the numbers are unfavorable, well the geologists can estimate some more reserves, the construction people can have second thoughts about the costs, etc. So, if those involved want it to fly, they can usually massage it into shape. They are often under pressure to make it work, whatever their personal judgment, because they may be bidding in a competitive situation where there is much more at stake than the specific project. Failure to participate may create a bad impression with the host government, which would have wider significance. The stock market too encourages companies to explore, naturally being ignorant of the real geological risks. If it is made to fly, the proposal is now blessed with a notional number showing it to be sufficiently profitable, and it passes up the management hierarchy, each level having less and less knowledge of the actual situation.

In reality, all that really sinks in at that stage is what the magic rate-of-return number is and what is left in the budget. The management desires a notional playing field and excludes local tax situations so that they can pretend to fairly compare the rate of return from investing in a refinery extension in Texas versus an exploration well in Norway. They thus fail to notice that 85% of the risk of the well in Norway is borne by the Norwegian taxpayer, who tacitly accepts that the cost of putting the well in the wrong place is deductible from taxable income. Companies with no taxable income in Norway could not take advantage of this tax break and soon withdrew from the game.

The one factor that really affects the economics, however they are conducted, is oil price. On that, the economists have little to contribute, because oil price has been largely politically contrived, although depletion does influence the long-term trend. They are not therefore in a good position to assess its distribution and accordingly cannot take into account the growing control of the resource by a few critical producers, which must surely influence the price more than ordinary economic factors.

Companies tend to have committees to assess future oil prices, mainly comprised of economists. They read the Wall Street Journal and consult Bloomberg, thinking in terms of supply and demand trends. Consequently, they normally come up with one or more bland scenarios, whereby oil price is above or below inflation by so many points. There is talk of the gentle ramp. Their record in forecasting has been abysmal.

But if all this seems rather negative and dismissive of the economist, in fairness we do admit that it is difficult to see how else centrally controlled global companies could run their affairs. Economic analysis does force those involved to think about all the aspects of the project. Moreover, companies clearly have near limitless opportunities to invest money: explore new areas, invest in different assets upstream or downstream, buy reserves or other companies, or invest in non-oil activities. So, they do need some yardstick by which to choose, and perhaps the economic analysis, in a very general way, does provide a comparison among prospects. The bland oil price assumptions are also understandable as it is difficult to plan for a crisis, even if crises are a normal fact of life. The system more or less helps the management avoid serious mistakes, even at the expense of not getting much right either. Above all, it helps managers and companies deal with their multiple responsibilities by allowing them to justify their decisions.

9.2   US Geological Survey

The USGS is a renowned agency dedicated to provide information on ecosystems, environment, natural hazards, natural resources, as well as impacts of climate and land-use change. Its Energy Resources Program has a division specialized in oil and gas resources making periodic assessments of the world's conventional oil and gas endowment since the oil shocks of the 1970s. The last comprehensive assessment was completed in 2000 and came to be known as the "USGS 2000" in the jargon of the oil debate—its official name is *World Petroleum Assessment 2000*. The USGS 2000 has been updated according to the priorities assigned by the USGS. Many agencies and organizations around the world, including oil companies and the IEA, use the data published by the USGS for their own forecasts and planning.

9.2.1   Different Criteria for North America

The USGS 2000 study gives estimates for undiscovered amounts of conventional oil, gas, and natural gas liquids (NGL), using a probabilistic approach. The USA, however, is treated differently from the rest of the world. First, oil and NGL are combined for the USA but distinguished elsewhere. Second, for the rest of the world, P95, P50, P5 (i.e., a low estimate which has a 95% chance of being realized, a "best guess" with a 50% chance, and a 5%, or high estimate), and mean cases are given by region, which are then aggregated using a Monte Carlo simulation—which is indeed the correct way to aggregate reserves. Curiously, for the USA, maxi (P95), mini (P5), and mean cases are quoted for the country as a whole, but the USA is not aggregated to the world total using a Monte Carlo procedure. If the non-US values are added using Monte Carlo, why is it not applied to the world when adding the USA? The failure to use a consistent method means that the assessment of P95 and P5 values for the world as a whole is fallacious.

9.2.2   Unjustified Discovery Rates

The proposed mean value of undiscovered liquids is 939 billion barrels (Gb) for the world, made up of 649 Gb of oil and 207 Gb of NGL outside the USA and 83 Gb for oil in the USA. It is claimed that the numbers relate to what may be discovered and added to the reserve base between 1996 and 2025, taking into account economic and technological factors. Such a claim of adding more than 50 billion barrels per year (Gb/a) is however very difficult to accept in relation to the past discovery trend, which has fallen from a peak in the 1960s to 10 Gb/a in the 1990s though with a slight recovery of approximately 13 Gb/a for the 2000s (Fig. 6.3). The USGS estimate implies a fourfold increase in discovery rate and reserve addition, for which no evidence is presented. Such an improvement in performance is in fact utterly implausible, given the great technological achievements of the industry over the past 20 years, the worldwide search, and the deliberate effort to find the largest remaining prospects.

9.2.3   Reserve Growth

In the USGS database, oil reserves "grow" due to the addition of previously undiscovered fields as well as the introduction of more efficient technology or the revision of past estimates (see Sect. 5.2.1 and Chap. 6). The USGS 2000 estimated 730 Gb for reserve growth, being made up of 612 Gb oil and 42 Gb NGL outside the USA, and 76 Gb for the USA.

First, in the former Soviet Union (FSU), there were 3,141 fields reported in 1997 but 3,930 fields reported in 2010, so 789 existing fields were missing in the first report. These older fields, if not considered appropriately, are accounted as "reserve growth." Second, in the USA, oil data has to meet the Securities and Exchange Commission (SEC) rules, while the rest of the world does not have to comply with them. These rules are designed to give certainty to the investor, not to assess the depletion of resources; the reserve estimates reported to SEC (proved reserves only, probable reserves are omitted) are usually lower than the real potential of a field, so the monetary return on investment is somehow guaranteed. Since the initial estimate was very low, after oil is extracted, it usually turns out that the reserves are larger than initially reported. These "extra" reserves give the false impression that actual reserves are growing (Fig. 6.2). The rest of the world reports proved and probable estimates because the industry has a greater need to know what the fields will actually deliver when they plan costly offshore facilities or pipelines to remote areas.

Thus, the huge "field growth" of the USA is clearly a reporting phenomenon, as only one out of every three barrels added over the past 20 years has come from new discoveries. While the cumulative new discoveries reported for 1990–2009 for USL48 were 5.36 Gb, the new discoveries plus discoveries in old fields added up to 18 Gb; meanwhile, the cumulative crude oil production was 37.5 Gb in the same period. Moreover, the USGS analysts extrapolated the model of growth of proved reserves in the old fields of the USA to the probable discoveries of the rest of the world. Even worse, they apply such a flawed method of assessment to present deep-water new fields. Schmoker (2000) uses the Midway-Sunset oil field as the best example of reserve growth. This field was discovered in 1894 and is a heavy oil field (13°API), classified by many as an unconventional field. Midway-Sunset peaked a century later, when production started falling in 1997. It is not the best example to use, as most new fields will not produce for a century before peaking. Jean Laherrère has stated that extrapolating US reserve growth to the rest of the world and also the deepwater fields even within the USA is unscientific.

In spite of these serious problems, due to the renown of the USGS, the World Petroleum Assessment was widely used and misused by other agencies—including the International Energy Agency—and is still cited in the debates concerning peak oil.

9.3   The International Energy Agency

After a period of dismissal, the International Energy Agency has begun to shift its ground to its previous assessment. The agency adopted Campbell and Laherrère's view in its *1998 World Energy Outlook* (WEO). The message they were sending was very clear:

This approach [...] indicates that a peaking of conventional oil production could occur between the years 2010 and 2020, depending on assumptions for the level of reserves (IEA 1998, p. 44).

However, in the 2002 edition, IEA described a different picture:

Resources of conventional crude oil and NGLs are adequate to meet the projected increase in demand to 2030, although new discoveries will be needed to renew reserves. [...] The approach used to generate these projections is described in Box 3.2 (IEA 2002, pp. 97–98).

When we look at Box 3.2, we read the following:

The oil supply projections in this Outlook are derived from aggregated projections of regional oil demand, as well as projections of production of conventional oil in non-OPEC countries and nonconventional oil worldwide. OPEC conventional oil production is *assumed to fill the gap* (IEA 2002, p. 95, emphasis added).

With this assumption, the IEA avoided having to produce a realistic estimate for OPEC production. As years passed, the agency found increasingly difficult to maintain this position. In 2008, the IEA admitted its previous assumptions did not match the reality of the oil fields (Monbiot 2008). A year later, the British press published an article based on the declarations of a senior official of the IEA, who revealed that the agency knew the predictions published in previous years were "nonsense" but fears about "panic in the financial markets," together with the pressure of "the Americans," prevented the IEA to lower the figures even more. A second source said it was a rule in the organization "not to anger the Americans" even though there was not as much oil in the world as the reports said (MacAlister 2009). In 2010, IEA finally admitted, that "conventional crude oil production" for the world had peaked in 2006. In Chap. 3 of the WEO 2010, titled "Oil Market Outlook: A Peak at the Future?," we read the following:

Almost half of the increase in proven reserves in recent years has come from revisions to estimates of reserves in fields already in production, rather than new discoveries. [...] in 2000–2009, discoveries replaced only one out of every two barrels produced –slightly less than in the 1990's (even though the amount of oil found increased marginally)– the reverse of what happened in the 1960's and 1970's, when discoveries far exceeded production (IEA 2010, p. 116).

When interviewed by the Australian media in 2011, Dr. Fatih Birol, chief economist of IEA, said that "global oil demand will increase substantially"; by contrast, on the production side, he said, "we think that the crude oil production has already peaked in 2006 [...] the existing fields are declining so sharply that, in order to stay where we are in terms of production levels, in the next 25 years, we have to find and develop four new Saudi Arabias." He added that one of the major conclusions of the WEO 2010 is that "the age of cheap oil is over" (Newby 2011). Compare Birol's comments and the article of Campbell and Laherrère in Scientific American vs the official pronouncements of IEA (including their 2012 pronouncement that the US will become an exporter of oil) and draw your own conclusions.

9.4   IHS Cambridge Energy Research Associates

IHS Cambridge Energy Research Associates (IHS CERA) is a well-known consultancy firm whose business is to deliver "critical knowledge and independent analysis on energy markets, geopolitics, industry trends and strategy" (IHS 2012). IHS and CERA were independent companies until 2004, when the former acquired the latter. Information Handling Services (IHS) was founded by Richard O'Brien in 1959, becoming specialized in databases during the 1980s. The expansion to oil consultancy is related to the acquisition of Petroconsultants S.A. in 1995, whose history goes back to the 1950s. Petroconsultants was an oil information service. Naturally, an oil company has every reason to track the activities of its competitors, which can have much commercial significance. In earlier days in the USA, they used to employ people known as "scouts" who would keep rigs under observation, sometimes with binoculars. They could, for example, count the stands of pipe being removed to figure out how deep the well was. Also they could hang around bars and talk to drillers having a beer. In the early days of the North Sea, oil companies placed observers on trawlers to watch rigs and if possible listen in to radio communications in the best traditions of scouting. It more or less amounted to what would be called industrial espionage today.

In the 1950s an American geologist named Harry Wassall worked for Gulf Oil and was transferred to Cuba, where he married a Cuban lady called Gladys. When Gulf Oil recalled him, he preferred to stay in Cuba and set up a little newsletter to report on oil activities on the island, later expanding it to cover Latin America. He appointed an agent in each country reporting on oil developments, including the location of new wildcats and the results. Much of it was not particularly confidential information.

When Fidel Castro came to power, he could no longer run this business from Cuba and moved to Spain, opening an office in Geneva to expand coverage around the world, naming it Petroconsultants. Over the years he built up a network of contacts, often comprising old oilmen with knowledge and experience of the particular country, who were able to build the database with continuity and trust. The major oil companies informally supported the endeavor as they preferred not to speak directly to each other but did want to know what each other was doing. They wanted good information and so they also gave it. In those days it was not a particularly sensitive matter. Also Petroconsultants was one of the first to apply computers to the database, and for a period, major oil companies found it convenient to subcontract their own databases to be managed in Geneva on a confidential basis. The company aged in parallel with its owner and became a rather charming old-fashioned organization staffed by old oilmen who had built long-term relationships and had the knowledge and background to assemble valid information.

Harry Wassall took an interest in the "peak oil" issue, seeing its wider significance. Petroconsultants read *The Golden Century of Oil*, published by Colin Campbell, which got much wrong due to unreliable public data (Campbell 1991). The firm invited Campbell and Laherrère to make a similar study using its database. The result of the study was eventually suppressed under pressure from an oil company, but Petroconsultants copublished Campbell's *The Coming Oil Crisis* and also encouraged both of them to write the *Scientific American* article.

Harry Wassall died in November 1995 and Petroconsultants was sold to IHS. The Geneva office has now put on a much more commercial basis, and most of the old staff left, taking with them their years of continuity, friendships, special relationships, and long experience. It accordingly became much more difficult to assemble privileged information, and the task itself became much harder because, with the growth of state companies and many small promotional companies, the major oil companies no longer dominated the business. In many cases it was not possible to do more than secure public information partly from the Internet and try to compile it as best as possible. CERA was an oil consultancy, run by Daniel Yergin, who received the Pulitzer Prize for his excellent book, *The Prize*, which describes the history of the oil industry from a business perspective in great detail (Yergin 2003). Yergin does not himself have oil industry experience, but the company could of course advise on oil developments and secure consultancies without having any particular detailed knowledge of the reserves of specific fields or countries. CERA was in turn acquired by IHS and now does have access to its database, for what it is worth.

However, IHS CERA has always forecasted optimistic scenarios about oil markets, and its executives have consistently argued that oil supply is ultimately driven by factors above the ground and not by any sort of geological constraint. In response to ASPO's critiques, CERA has also argued that a long "undulating plateau" extending over "several decades" is more likely pattern than a peak in oil production (IHS 2009a); this plateau would start, in the third or fourth decade of the century. We would like to point out that neither Hubbert nor Campbell, Laherrère, or ourselves have ever said that geology is the sole driver of oil supply; rather, we believe that there are limits of different kinds to oil supplies, and given the discovery trend of the last decades, together with the decline in producing fields and the state of technology in the oil industry, it is not likely that oil supplies will reach a higher level in the following decades for geological reasons. In addition aboveground conditions, wars, boycotts, political manipulations, and economics can constrain (or possibly enhance) that limitation.

Since CERA is a private consultancy, their predictions are not accountable; when they release a so-called private report, it means that the report can be bought by anyone for US $2,500. The data files used in the report are also "private" rather than being audited or refereed like the data in scientific articles. Nevertheless, ASPO and other observers have kept track of their figures. In 2002, they predicted that North American natural gas production would increase 15% by 2010. In reality the production remained flat until 2008. In 2003, CERA estimated that oil prices would fall to low or mid $20s, while they actually remained above 30 US dollars. In 2004, they said oil prices would be in the range of upper $20s to low $30s thru 2005, but the prices climbed to $65. Then, in 2005, their forecast was a decline towards $40 as 2007–2008 neared, yet again, the price stayed in the mid $50s. In 2007, they predicted prices for the next year as low $60s, but prices reached $90 (Energy Bulletin

2008, see also Brown 2011). In 2008, a group of businessmen and energy experts, including Jean Laherrère, issued a $100,000 wager against the forecast that CERA published in June 2007. The forecast said world oil production capacity would reach 112 million barrels per day (Mb/d) by 2017 (IHS 2008a). That figure would imply roughly 107 Mb/d of actual production, a number that could be easily verified. CERA never answered the wager (Andrews 2008).

Since 2008, as actual oil production has remained flat, IHS CERA has been claiming that demand for oil products has peaked due to high oil prices. In 2008, they stated that gasoline demand had peaked a year before in the USA (IHS 2008b). In 2009, CERA accepted that "peak oil is here," but not because of any underground constraints, but because the oil demand had reached its limit; they said that demand for oil in OECD countries was not likely to return to its 2005 high and that "aboveground drivers" would be crucial to meet growing demand from non-OECD countries (IHS 2009b). Nevertheless, CERA's statements about peak demand conveniently forget the geological causes of the historically highest oil prices that we have been enduring in the last years. These high prices might nurture strong investments in lower-quality resources, such as tar sands and shale oil, whose extraction and environmental costs are larger. However, these costs are not factored into the economic calculations of CERA.

In conclusion, given the critical importance of oil to modern society and the unresolved issues and controversy swirling around "peak oil," it is remarkable that governments do not insist on some kind of solid, technical database. Instead we have a series of very different assessments published by private or public entities that summarize information coming from multiple sources of unknown veracity. Very often the estimates given are a function of the political or economic perspective of the supplier. As scientists used to substantiating values, open analysis, examination of information sources, peer review and, ideally, open discussion of differences, we find the situation amazing.

References

Andrews S (2008) About the $100,000 CERA bet: ASPO-USA: Association for the Study of Peak Oil and gas. ASPO USA. http://aspousa.org/about-the-100000-cera-bet/. Accessed 30 Jul 2012

Brown JJ (2011) Daniel Yergin massively reduced his energy estimates. Energy Bulletin. http://www.energybulletin.net/stories/2011-10-24/daniel-yergin-massively-reduced-his-energy-estimates

Campbell CJ (1991) The golden century of oil, 1950–2050: the depletion of a resource. Kluwer Academic, Dordrecht/Boston

Energy Bulletin (2008) Group bets $100,000 against CERA supply forecast. Energy Bulletin. http://www.energybulletin.net/node/39973. Accessed 30 Jul 2012

IHS (2008a) No evidence of precipitous fall on horizon for world oil production: global 4.5% decline rate means no near-term peak: CERA/IHS study. IHS Online Pressroom. http://press.ihs.com/press-release/corporate-financial/no-evidence-precipitous-fall-horizon-world-oil-production-global-4. Accessed 30 Jul 2012

IHS (2008b) CERA: "Peak Demand"—U.S. gasoline demand likely peaked in 2007. IHS Online Pressroom. http://press.ihs.com/press-release/energy-power/cera-%E2%80%9Cpeak-demand%E2%80%9D-us-gasoline-demand-likely-peaked-2007. Accessed 30 Jul 2012

IHS (2009a) IHS CERA: Oil supply set to grow through 2030 with no peak evident. IHS Online Pressroom. http://press.ihs.com/press-release/energy-power/ihs-cera-oil-supply-set-grow-through-2030-no-peak-evident. Accessed 30 Jul 2012

IHS (2009b) Oil demand from developed countries has peaked. IHS Online Pressroom. http://press.ihs.com/press-release/energy-power/oil-demand-developed-countries-has-peaked. Accessed 30 Jul 2012

IHS (2012) Energy strategy: IHS CERA. http://www.ihs.com/products/cera/index.aspx. Accessed 30 Jul 2012

International Energy Agency (1998) World energy outlook 1998. Organisation for Economic Co-operation and Development; Distributed by OECD Publications and Information Center, Paris; Washington, DC

International Energy Agency (2002) World energy outlook 2002. Organisation for Economic Co-operation and Development; Distributed by OECD Publications and Information Center, Paris; Washington, DC

International Energy Agency (2010) World energy outlook 2010. Organisation for Economic Co-operation and Development; Distributed by OECD Publications and Information Center, Paris; Washington, DC

Macalister T (2009) Key oil figures were distorted by US pressure, says whistleblower. The Guardian.http://www.guardian.co.uk/environment/2009/nov/09/peak-oil-international-energy-agency. Accessed 9 Jul 2012

Monbiot G (2008) When will the oil run out? The Guardian. http://www.guardian.co.uk/business/2008/dec/15/oil-peak-energy-iea. Accessed 30 Jul 2012

Newby J (2011) Oil crunch. In: Catalyst: ABC TV Science. http://www.abc.net.au/catalyst/stories/3201781.htm. Accessed 30 Jul 2012

Schmoker, JW (2000). Reserve Growth Effects on Estimates of Oil and Natural Gas Resources (Fact Sheet No. 119-00). US Geological Survey. Retrieved from http://pubs.usgs.gov/factsheet/fs119-00/ . Accessed 12 October 2012

Yergin D (2003) The prize: the epic quest for oil, money, and power, 1st trade paperback edn. Free Press, New York

Bibliography

Section 9.1 is based on the following sources:

Campbell CJ (2005) Oil crisis. Multi-Science Publishing Co. Ltd, Brentwood (Chapter 10)

Section 9.2 is based on the following sources:

Laherrère JH (2000) Is the USGS 2000 assessment reliable? Cyberconference by the World Energy Council. http://www.oilcrisis.com/laherrere/usgs2000/

Laherrère JH (2001) Estimates of Oil Reserves. In: International Energy Workshop, Laxenburg, Austria. http://webarchive.iiasa.ac.at/Research/ECS/IEW2001/pdffiles/Papers/Laherrere-long.pdf

Chapter 10
Conclusions

Charles Hall is old enough to remember the shock and exhilaration of coming across the work of M. King Hubbert in the National Academy of Sciences book *Energy and Man* in 1969 while browsing the bookstore of the University of North Carolina. He was an environmentally sensitive graduate student at the time and had just come across Jay Forrester's original "Limits to Growth" article in his father's MIT alumni magazine. He found the concept that all that he saw around him, good and bad, including his tremendous mobility and the availability of graduate education for many, including himself, as well as the mindless development of his beautiful coastal home town and the absurd Vietnam war where his friends were serving and being killed, depended on the availability of cheap petroleum which might not last his lifetime. So while his immediate focus was on systems ecology applied to energy use in streams and fish migration, he began to understand (greatly encouraged by his graduate advisor Howard T. Odum) that energy principles applied equally to human endeavors. A postdoc at Brookhaven and Oak Ridge National Laboratories, whose main focus was on nuclear processes, did little to dispel his realization about the importance of energy to most things humans were doing at that time.

Carlos Ramírez-Pascualli became interested in economics after having graduated from Mexico's premiere university in 2001. He realized there was a large mismatch between the excellent preparation he received in the National University and the jobs available in Mexico for young engineers like him. Many professionals remained unemployed—or subemployed at best—despite the urgent need to develop technology and infrastructure at all levels. To understand the nature of the problem, he decided to enter one of the top graduate programs in economics in Mexico. Coming from a discipline deeply rooted in empirical knowledge, where the common practice is not to take theory as reality and whose major concern is to solve problems with the least possible amount of resources, it was striking for him to learn that "efficiency" in economics depended on incommensurable, individualistic valuations. The concept was devoid of any attempt to understand, measure, or account for any biophysical variable, especially the nonrenewable energy required to feed the industrial processes that transform raw materials into goods and services.

C.A.S. Hall and C.A. Ramírez-Pascualli, *The First Half of the Age of Oil: An Exploration of the Work of Colin Campbell and Jean Laherrère*, SpringerBriefs in Energy, DOI 10.1007/978-1-4614-6064-0_10, © Springer Science+Business Media New York 2013

This severe shortcoming had led Mexico (or at least, had provided the justification) to export roughly half its oil production at the time and to reduce its industrial sector significantly. Both processes were clearly against the experience of a once-poor country that had maintained high rates of economic growth for 30 years (1940–1970) using oil as a physical asset for industrialization instead of a collateral to borrow (1976–1982) and a source of foreign currency to repay external debt (1982–2003). It was clear that Mexico had squandered its energetic wealth, but the tools of economic theory were useless to understand the general situation of the country and may have exacerbated the problem.

In the USA, Hall saw an explosion of interest in peak oil in the 1970s and early 1980s as the country was subject to two "energy crises" and the price of oil shot up from \$3.50 to \$60 a barrel. He had at that time two very special undergraduate students, Robert Kaufmann and Cutler Cleveland, with whom he published a series of papers in *Science* magazine and elsewhere, and produced a book that showed clearly that oil was becoming harder to get and that all kinds of basic economic concepts for the USA (e.g., production of goods and services, labor productivity, inflation) could be predicted over 100 years from energy alone with R^2s of 0.96 or greater, and that increased drilling did not lead to increased finding or production of oil.

But by 1986, when the book came out, the national interest in energy, once enormous, had disappeared. Gasoline and heating prices came down, general inflation had caught up with energy prices, and few paid much attention. Intellectually, economists, who had been arguing against any limits to growth and the importance of any particular resource to economies in general, appeared to have won the day as the higher oil prices encouraged the development of oil fields that had been found before the 1980s but that had been previously too expensive to develop when oil was sold at \$3.50 a barrel. Writers such as New York Times' John Tierney took the side of the economists who had argued that market economics would resolve any resource problem. For the most part, public and academic interest in energy and its effects on the economy simply died, except for a small cadre of scientists nearly all of whom undertook their research "after retirement, on weekends or pro bono." Any national funding for energy-related science tended to focus on the development of some kind of new technology, while society as a whole remained almost exactly as dependent, proportionately, on oil, gas, and coal as it had been in 1973, even while the amounts used increased enormously. Energy was essentially absent from any discussion in scientific circles through the 1990s.

Campbell and Laherrère reignited the interest in peak oil. Hall noticed this first with their 1998 publication in the *Scientific American* of "The End of Cheap Oil." At that point he had no idea who they were, but he thought the article was wonderful. Someone had picked up Hubbert's baton, and they knew what they were doing! This and their related efforts led to a second explosion of broad interest in oil, energy, and its relation to the economy (although of course most people still did not have a clue). As such, we consider that these two geologists have been among the most important scientists of the past two decades. While they are both too humble to accept this designation and are probably embarrassed to read these words, we cannot help but arrive at this conclusion, from reading and editing the chapters of this book.

Where did these guys come from? What is their particular background that allows them, rather than someone else, to take the new lead in understanding and promulgating this approach to oil? In this book we have summarized the answers to these questions. First of all, they were extremely respectable petroleum geologists with excellent training in their nations' top universities (Campbell received his degree from Oxford University and Laherrère from the *École Polytechnique*). Second, they joined the oil industry at a very early stage, so they have large field and analytical experience in finding and managing oil fields around the world. They witnessed the development of multiple exploration and production technologies, from surface exploration in Colombia to seismic techniques in Algeria, from securing positions in the Sea of Barents to the publication of several technical manuals. Third, they both reached privileged positions inside the oil industry, first as high-level directors in important oil companies and, later, as consultants associated to a renowned firm. Campbell was exploration manager for Aran, in Dublin, and later for Amoco, in Norway, and also executive vice president for Fina, in Norway too. Laherrère was deputy exploration manager for TOTAL, president of the Exploration Commission of the *Comité des Techniciens, Union Française de l'Industrie Pétrolière,* and director of *Compagnie Génerale de Geophysique,* Petrosystems, and other TOTAL subsidiaries. Their careers gave them a firsthand experience with the corporate bureaucracy, the information management, and the financial practices in the international oil industry. After retirement, they became consultants with a network of connections that granted them access to private data. Fourth, by contrast to the conventional environmentalist position, often portrayed as extremely pessimistic, Campbell and Laherrère have addressed the study of oil and gas depletion using geological and quantitative analyses, and they have done so without having any vested interest, whether economic or political; they are simply guided by the data. Last but certainly not least, they have been interested in the social and political ramifications of oil and gas depletion, something that the oil industry does not exactly encourage. They independently realized the importance of the larger view, that not only was oil becoming harder and harder to find in significant quantities (all oilmen know that) but what this difficulty to find oil meant collectively over all oil fields and all oil-producing nations, and what it would mean ultimately for the industrial civilization. Thus, the work of Colin Campbell and Jean Laherrère is an essential reading to understand oil depletion and its implications. From their writings and presentations, complemented with a little of our own research, we have come to appreciate the complex connections between oil, money, and the industrial civilization. These connections can be best understood when we integrate geological, physical, and chemical knowledge into the study of economic systems.

If we take a biogeological perspective, we can easily see that our current lifestyle is a fragile achievement indeed, a point that Campbell and Laherrère have emphasized throughout their numerous writings and presentations. The history of modern society started a few centuries ago, yet the industrial mode of production that has supported it depends to a large extent on the combustion of fossil hydrocarbons whose generation and accumulation requires millions of years. The formation of fossil hydrocarbons requires very specific geological and biological conditions that tie our civilization to the history of our planet and the life in it. This history requires an even larger scale, ranging to billions of years. From this biogeological perspective, our civilization, with all its technology, affluence, and comfort, seems more of a transient phenomenon than a robust construction built upon solid foundations. Following the tradition of Hubbert, Campbell, Laherrère, and the peakists, we the people who live in the modern society need to study and discuss the issues related to the depletion of fossil fuels, bearing in mind the short timescale of our life and our civilization in relation to the biogeological processes upon which it is based.

To illustrate the timescales at stake as well as the connections between culture and resources, Campbell studied the biological and cultural evolution of *Homo sapiens*; in this book, we have updated and complemented his initial efforts in the light of recent anthropological and archeological discoveries. In this sense, while the formation of fossil fuels started hundreds of millions of years ago, the hominid line that led to humans appeared in Africa only four millions years ago. Furthermore, it took 3.8 of these four million years for the *H. sapiens* to evolve, plus some extra 140,000 years for them to migrate out of Africa. The elements of civilization started appearing only ten thousand years ago with sedentism and agriculture, followed later by copper and gold metallurgy (seven thousand years ago), and then bronze metallurgy (three and a half thousand years ago). Cultural features, such as religion, accounting, and coinage, started unfolding together with these elements. All these techno-cultural factors contributed to the appearance of a culture that started flourishing only a few hundreds of years ago in Europe. This culture emerged around steel metallurgy and the energy accumulated for millions of years in fossil hydrocarbons, the two basic components of the engines run by steam expansion or by combustion of liquid fuels. The same as agriculture or metallurgy, engines brought changes to the social and economic organization, leading to a production system that we now call capitalism. This system successfully aligned the technological and social developments with a culture of "progress" and "mastery of nature." In other words, the organization of labor around engines fueled with fossil hydrocarbons became politically and culturally acceptable in terms of social "progress," although not without social struggles. In some cases, these conflicts forced populations to migrate overseas in search for better life conditions, exporting the Western technology and culture to different parts of the world. Campbell has been especially careful to indicate that much of this migration was a result of desperation rather than a question of free will.

Population growth and migration are other issues that Campbell and Laherrère have linked to the exploitation of hydrocarbons, especially oil. While coal initiated the process of industrialization in Europe, oil arrived in the second half of the nineteenth century, in the midst of an industrial world hungry for fuels to support its new technologies and its economic expansion. The coal economy had allowed the population numbers to double, and the system organized around oil made possible a sixfold expansion of the population in only 150 years, mainly through the extension of longevity. These changes resulted eventually in the unforeseen fertility distributions and some of the migration patterns that we witness today. In general, this "progress" was attributed to the allocation mechanisms developed in the industrialized societies, which were conceptualized in terms of a market that coordinated the forces of supply and demand. According to this view, globalization would be the outcome of companies seeking to open new markets. This picture, however, did not acknowledge that the organization of a society and its technology are determined not only by culture but also by the physical characteristics of the resources it has come to exploit. As Campbell and Laherrère have pointed out in many places, awareness about the biophysical processes involved in our social organization has been relegated to the background, when in fact the emergence of the capitalistic system, the industrial era, and globalization was made possible by the unique physical properties of fossilized hydrocarbons. These properties are the result of processes that can be measured only in the geological timescale.

Campbell has also explored the link between oil and economic value from a chemical perspective. Hydrocarbons have a very large heat value, and hence economic utility, due to their chemical composition. As a chemical element, carbon has an outstanding ability to form stable bonds with itself and other elements at standard conditions; however, all its compounds are combustible if an ignition source, high temperatures, high oxygen concentrations, or any combination of these are present. The other component, hydrogen, forms combustible compounds too, forming bonds that release more energy than carbon bonds do. Hence, the combustion of hydrocarbons liberates a great amount of heat into the surroundings. Carbon compounds containing oxygen, such as carbohydrates and alcohols, are combustible too, but do not release as much heat because their oxygen atoms cannot be oxidized and, hence, cannot liberate energy. This is why neither biomass nor biofuels can be engineered to the point where they could contain more energy than coal, oil, and natural gas.

Some may object that oil was already there when the great civilizations of antiquity were flourishing, yet only the Western civilization was "ingenious" enough to exploit the substance. While this might be true, there is more to the story; Campbell has illustrated that the major difference between the West and other civilizations that knew oil was not the technology to obtain oil, but the evolution of the demand for the substance. Oil was known since antiquity, and some of the extractive technologies still in use in nineteenth century Europe and North America can be traced back many centuries to China. The Chinese and other cultures used oil for ritual purposes, medicinal treatments, heating, and illumination. Except for the West, no other civilization used oil to run engines, which are indeed a Western feature. In the beginning, the West was not different, the early oil industry unfolded rapidly in the Pennsylvanian Appalachians and at the shores of the Caspian Sea speared by the urgency for a new illuminant to replace whale oil. Some of the large oil companies, such as Standard Oil, Shell, Gulf, and Texaco, were born in the era when oil was principally an illuminant. For example, Standard Oil was established at the time when Carl Benz was inventing the automobile. Indeed, the internal combustion engine gained importance as the political tensions between England and Germany increased, becoming crucial in World War I.

Campbell correctly identified this conflagration as one of the important tipping points in the history of the oil industry. It was probably the first and clearest example of the geopolitical implications petroleum would have. Even before the beginning of hostilities, British and Germans were engaged in an arms race based on the speed of their fleets. Steam power could not compete against oil-derived fuels. Since Europe is not a prolific oil province, both British and Germans had to secure their respective oil supplies outside their neighborhood, and both were led to the Middle East. The war itself started with horses and finished with oil-fueled tanks and air fighters. Oil not only was crucial during the war, it also played a large role in the aftermath. The Middle East, formerly under the control of the Turkish Empire, came under the sphere of influence of the Allies, who were especially interested in its oil reserves. The political division that we see today in the Middle East is an outcome of this influence. Thus oil became not only an important asset for the European and North American industries but also a source of political struggle for the coming decades, including the years of World War II.

The early postwar decades (1950–1970) were a period of bonanza for the oil industry. Global oil production increased 4.5 times during this period. However, the trend in reserves was already showing signs of decline. Among the first geologists to point out that imbalance and to address the social consequences of the process was M. King Hubbert. In a famous lecture given in 1956, he presented some calculations indicating a peak in US oil production occurring between 1965 and 1970, followed by a decline. His forecast was pretty accurate in some sense: within the assumptions and data he used, Hubbert's prediction was exact: oil production in the USA did peak in 1970 for the reasons he presented in his lecture. Supply problems in the USA started to gain importance only after the peak, during the oil shocks of the 1970s. However, the model was misleading after the occurrence of the US peak. Hubbert did say that, after the peak, the production curve "must decline at a rate comparable to its earlier rate of growth," something that did not happen due largely to the discovery of Prudhoe Bay, Alaska, in 1968. From our vantage point, we may say today that Hubbert was too pessimistic with respect to the capacity of the industry to find more oil in general. Nevertheless, the challenge he launched to the projections of everlasting growth in the oil industry would remain dormant but alive for the decades to come.

Afterwards, few authors tried to reproduce Hubbert's methods at a larger scale until Colin Campbell and Jean Laherrère joined forces to undertake a comprehensive study along these lines. This study was intended for sale inside the oil industry, but due to pressures coming from some oil companies, the study was not disseminated to the industry at large. Instead, its basic results were published in "The End of Cheap Oil." This article captured the attention of companies, organizations, politicians, private analysts, scholars, and activists. Unfortunately, the data used for this analysis is private so the study cannot be verified by an independent source. However, despite all the uncertainties involved, there are ways to verify the forecast of Campbell and Laherrère using data available to the public. Jean Laherrère has been especially careful to document these trends thoroughly. For example, in the last decade, oil production did not grow at a significant rate as compared to the previous decades despite the growth of global demand and the consequent price increase. In fact, since 2005 oil production (all liquids included) has been bouncing around a plateau at 86 Mb/d plus or minus 2 Mb/d. Prices have increased since 2001, except for 2 years: 2009—the year of the international economic crisis—and probably 2012 too. This means that during a decade, the oil industry was unable or unwilling to provide *cheap* oil. Even if production grows and prices fall this year as expected, these variations will represent around 2% relative to the past year in both cases: total production staying below 90 Mb/d and prices staying above 90 dollars. It seems adventurous to forecast variations like the ones that happened during the aftermath of the oil shocks of the 1970s: in the mid 1980s, oil production grew at low prices until the year 2000.

"The End of Cheap Oil" was different from other "pessimistic" predictions, not only because of the special qualifications of its authors but also because it served as a solid platform for further analysis and discussion. The article was the initial spark that ignited a movement that Campbell and Laherrère led, side by side with other analysts, scholars, and activists: the Association for the Study of Peak Oil and Gas, commonly known as ASPO. The association began as an informal network of scientists trying to study the patterns and impact of oil depletion, and it has remained like that so far. Nevertheless, private companies and public institutions all around the world have acknowledged its voice.

ASPO has been nurtured by a wide diversity of perspectives, from geology to sociology. However, as this book has documented briefly, it is clear that the unique position capitalized by Campbell and Laherrère—in terms of training and experience in many different spheres of the oil industry—as well as the thesis brilliantly embodied in "The End of Cheap Oil" were essential for the formation and consolidation of ASPO, a project that could have easily been marginalized as many other valuable efforts have been. Today, ASPO has more than thirty international chapters from all the continents. Its members have produced numerous articles, books, films, and even a Depletion Protocol related to "peak oil." However, additional efforts are required to analyze and communicate the significance of oil and gas depletion because the general public and the press are much more influenced by quite small new developments such as the Bakken oil.

We believe the term "peak oil" introduced by Campbell has been very useful to communicate the issue of a nongrowing oil supply. One problem, however, is that some use of the term leads to catastrophic scenarios and undue speculations. While we believe that the gap between actual supply and demand projections is likely to increase, with possible consequences for the economic system, we certainly do not view catastrophe as inevitable. As Hubbert stated in his lecture in 1956 and Jean Laherrère has insisted elsewhere, there can be several peaks before a long-term decline starts. In fact, a peak is only an analytical artifice that works only in terms of annual production. Quarterly and monthly plots exhibit a great degree of variability with multiple peaks appearing over time. Laherrère has also pointed out that a bumpy plateau seems to be a more likely scenario than a sharp peak. So far, it seems that his insight is accurate.

In the last chapter we presented a brief overview of the decision-making process inside the oil companies and a sample of the different positions and critiques that Campbell, Laherrère, and the other "peakists" had been facing throughout the years. Most of these critiques have come from economists, financiers, and business analysts, inside and outside the oil industry. We tried to take these arguments as seriously as possible, and while we have found some part of them to be reasonable to some extent, we also find them systematically incomplete and, hence, misleading. Inside the oil industry, the dynamics of the business, at least in the experience of Campbell, has created an atmosphere of professional competition that results in the optimistic assessment of projects. The economic analysis inside the industry should be interpreted more as a tool to avoid serious mistakes rather than as a calibrated instrument to measure the status of global resources. Besides, the political pressure of the governments, together with the financial pressure of the markets and overconfidence in technology has influenced the companies to undertake optimistic evaluations of reserves. Thus, as Campbell found out in his first studies, the information available to the public is largely unreliable.

As any other business, the oil companies manage their own information, selecting what data is most suitable for their interests. The reports they emit to official agencies should be understood as a strategic statement and not as part of their commitment to society. We are not denouncing the strategy of the private companies as illegal or unethical (but we are not approving it either); we are only pointing out that their first priority is to make profit, not to inform society. To think otherwise is naïve or perverse. With regard to the national oil companies, their situation is not better. Many of them are used to fulfill the projects of the state or government in turn and usually are not a reliable source of information. The extensive analysis that Jean Laherrère has done over the evolution of official reserves along the years illustrates this point in a clear way.

We also believe the discussion has to move beyond the particular shape of the oil production curve. The existence of a single or multiple peaks, a plateau, or any other possible shape, is secondary. Again, what is at stake is the high cost of the investments our society has to make in order to obtain oil or any other kind of energy. Supposing OPEC or the Saudis were inflating the price of oil artificially, they do so because they can, and they can do it because the other companies are not able to extract enough cheap oil to compete with them despite all the technological improvements they may have achieved. Therefore, the geology of Venezuela and the Middle East has been much more important than technology itself.

Finally, this book analyzed the position of some influential institutions through their own publications and other media reports. The study done by the USGS in 2000 was a valuable effort, but it needs to be updated, and the assessment techniques need to be significantly improved. The International Energy Agency (IEA) was trying to send precautionary signals back in 1998, following the arguments of Campbell and Laherrère. In 2002, it entered a period of denial, when it assumed simplistically that OPEC producers would fulfill the demand of the West. In the last few years, the IEA has been trying to return to its former position, acknowledging that the end of cheap oil is here although powerful interests seem to push them into unduly optimistic predictions as of November 2012. Even IHS CERA, a consultancy well known for its polemic against ASPO and its opposition to the idea of an oil supply driven by geological constraints, has recently admitted that "peak oil demand" has finally arrived to the West. In other words, they think demand will not increase more at the current prices, which happen to be historically high. Can we expect prices to come down? What are the drivers for the high prices? Again, even if OPEC or the taxation regime were the leading drivers behind these high prices, the most likely conclusion is that all the technology developed in the oil industry is not able to compete against the cheap Middle-Eastern oil.

We all know that humankind will not use oil indefinitely; therefore, there will be a period in human history when oil production starts declining. The debate is whether this decline will happen because we will have developed a better source of energy—or at least a decent replacement—or because we will not be able to extract as much oil as before. In this book, we have tried to show that the bumpy plateau we are looking at today is not the sole result of OPEC's policies or of a bad financial environment, but the result of the end of cheap oil, a geologic event whose social and political consequences remain to be seen, a concept that we owe to the two men who inspired this volume and whose work and dedication we have tried to honor in these pages. Our recognition and sincere gratitude to them, Colin Campbell and Jean Laherrère.

Index

A
Africa, 9, 14, 45, 67, 79, 118
Agriculture, 10, 12, 97, 118
Alaska, 44, 45, 47, 52, 56, 83, 84, 120
Alcohols, 22, 119
Aleklett, K., 3, 93, 95, 96
Algae, 7, 8, 24
Alkanes, 22, 24
Appalachian boom, 31–34
Arabian American Oil Company (ARAMCO),
40, 44, 59, 93
Aromatic hydrocarbons, 22–25
Association for the study of peak oil (ASPO),
1, 3, 55, 59, 82, 89–99, 101, 111,
121, 123
Atlantic Richfield Company (ARCO), 44, 45

B
Bacteria, 8
Baku, 29, 31, 35, 36, 41
Biofuels, 22, 119
Britain, United Kingdom, 11, 12, 15, 37–40,
44, 46, 65, 69
British Petroleum (BP), 36, 38–41, 44, 45, 55,
74, 84, 86, 90, 102
Bronze Age, 10, 11, 20

C
Cambrian, 6, 11
Cambridge energy research associates
(CERA), 110–112, 123
Campbell,
Canada, 26, 35, 41, 46, 66, 80, 94
Carbohydrates, 19, 21, 22, 119

Carbon, 19–24, 119
Caspian Sea, 2, 29, 119
CERA. *See* Cambridge energy research
associates (CERA)
Chemical bonds, 20
Chevron, 33, 34, 39–41
China, Chinese, 11, 14, 29, 30, 59, 61, 96, 119
Civilization, 1, 2, 5–16, 19, 21, 29, 117–119
Colombia, 15, 35, 41, 86, 90, 95, 117
Combustion, 12, 21, 39, 117–119
Conventional oil, 56, 57, 62, 63, 65, 66, 68,
106, 107, 109
Curve, 47–53, 57, 60, 63, 69, 74–82, 89, 122
Cycloalkanes, 23–24

D
Data, 1, 3, 7, 48, 55–57, 71, 72, 74–76, 82–86,
93, 101, 102, 107, 108, 110, 111,
117, 120, 122
Deepwater, 44, 46, 65, 75, 76, 108
Density, 21, 22, 25, 26
Depletion, 1, 3, 5, 14, 16, 31–33, 46–48, 50,
52, 53, 55, 73, 91–95, 97–99, 101,
106, 108, 117, 118, 121
Discount rate, 105
Discoveries, 33, 58, 62, 78–82
Drake, E.L., 30–32
Drill, drilling, 30, 31, 35–37, 45, 46, 55, 61,
66, 83, 90, 91, 99, 103, 105, 116

E
Earth, 2, 5–7, 11, 20, 30, 58, 93
Economy, economics, 5, 14, 16, 68, 95, 96, 99,
116, 118